SIMPLE STATISTICAL METHODS FOR SOFTWARE ENGINEERING

DATA AND PATTERNS

SIMPLE STATISTICAL METHODS FOR SOFTWARE ENGINEERING

DATA AND PATTERNS

C. Ravindranath Pandian
Murali Kumar S K

CRC Press
Taylor & Francis Group
Boca Raton London New York

CRC Press is an imprint of the
Taylor & Francis Group, an **informa** business

AN AUERBACH BOOK

CRC Press
Taylor & Francis Group
6000 Broken Sound Parkway NW, Suite 300
Boca Raton, FL 33487-2742

First issued in paperback 2019

© 2015 by Taylor & Francis Group, LLC
CRC Press is an imprint of Taylor & Francis Group, an Informa business

No claim to original U.S. Government works

ISBN-13: 978-1-4398-1661-5 (hbk)
ISBN-13: 978-0-367-37764-9 (pbk)

Contents

SECTION II METRICS

SECTION IV TAILED DISTRIBUTIONS

Preface

This book is a tribute to great Statisticians, scholars, and teachers whose ideas are quoted throughout this book in various contexts. These pearls of wisdom have helped us to connect our book with the evolution of science, knowledge and engineering. Eventhough there are many books on statistics, there are few dedicated to the application of statistical methods to software engineering. Pure textbooks provide scholarly treatment, whereas practitioners need basic understanding and application knowledge. Very few statistical books provide application knowledge to software engineers. We have been working toward bridging this gap for about two decades and have come out with the current book.

Statistical methods are often discussed in the context of six sigma, Capability Maturity Model Integrated (CMMI), establishing capability baselines, and constructing process performance models. Driven by CMMI auditors, such practices have become rituals that rely heavily on automated statistical packages, which are rarely well understood. We have been promoting excel-based solution to statistics and have presented practical solutions, such as those achieved in this book.

> Statistics is the grammar of science.
>
> **Karl Pearson**

We also realize that sophisticated statistics is not the ideal approach to solve problems. Simpler techniques provide easy solutions that connect with the intuition of problem solvers. Although sophisticated techniques sound impressive but merely academic, simpler techniques are flexible and can easily penetrate to the root of the problem. In this book, we have consciously selected simpler tools. We have also simplified several standard techniques.

The techniques presented in this book appear to us as a minimum set of intellectual tools for software engineers and managers. True software engineering can happen only when data are collected and these statistical methods are used. Moreover, the statistical management of processes is possible only when managers master these techniques.

Learning these techniques in the context of software engineering will certainly help budding engineers and fresh recruits. The examples provided in this book will provide a deep insight into software engineering and management.

This book can be used extensively as a guidebook for training software engineers and managers at different levels. It will be a very valuable asset in the hands of quality professionals who collect data and create models.

This book also exposes practical software engineering problems and solutions to aspiring engineering graduates and make them industry ready.

Generally, this book is a guide for professionals to think objectively with data. It will help them to mine data and extract meanings. Some of the techniques provided in the book are excellent prediction tools, which would give foresight to those who apply them.

MATLAB® is a registered trademark of The MathWorks, Inc. For product information, please contact:

The MathWorks, Inc.
3 Apple Hill Drive
Natick, MA 01760-2098 USA
Tel: 508 647 7000
Fax: 508-647-7001
E-mail: info@mathworks.com
Web: www.mathworks.com

Acknowledgment

This book would not have become a reality without fruitful feedback from several software professionals, quality managers, and project managers who have taken our training and consultancy services. We also acknowledge Software Process Improvement Network (SPIN) for presenting some of these concepts through various SPIN chapters in India in an attempt to propagate these methods. All the SPIN coordinators we interacted with have provided excellent application suggestions.

We thank those organizations who have shared their metric problems with us for analysis and resolution. They eventually provided us research opportunities that helped us gain deeper knowledge. We also thank many research scholars who have interacted with us and taken our research support in the context of data mining and artificial neural network.

We thank the professors and correspondents of many colleges in India for helping us interact with students. We also thank Project Management Institute (PMI) chapters and project management institutes who gave us opportunities to present quantitative techniques to managers.

Rathna and Samuel helped by offering a wonderful review and criticism of Chapter 8. Swaminathan contributed to Chapter 21 by reviewing the chapter and making valuable suggestions. Shanti Harry helped us with references and suggested readings. We thank all these well wishers.

Finally, we thank Mr. John Wyzalek who provided moral support and editorial help. He made serious decisions about the scope of this book and helped us make a tough decision to leave some chapters for the future and focus on the few we have selected for this publication.

Introduction

The book contains four sections. In the first section, we present facts about data. In the second section, we recapitulate metrics. In the third section, we cover basic laws of probability. In the fourth section, we present special data patterns in the form of tailed mathematical distributions.

We are addressing development metrics, maintenance metrics, test metrics, and agile metrics in separate chapters, paying special attention to the specific problems in each domain. We also cover the construction of key performance indicators from metrics.

We also present elementary statistics to understand key characteristics of data: central tendency and dispersion in two separate chapters. The great contribution from Tukey in creating a five-point summary of data and the box plot is presented in the special chapter.

In Chapter 10, we introduce pattern extraction using histogram. These patterns are empirical in nature and are priceless in their capability to show reality as it is. Going forward, these empirical patterns are translated into mathematical patterns in individual chapters in terms of statistical distributions. Examples are provided in each chapter to understand and apply these patterns.

Each chapter is illustrated with graphs. Tables are used to present data where necessary. Equations are well annotated. Box stories are used to present interesting anecdotes. In particular, brief notes are presented about original inventors of ideas. Each chapter contains references on key subjects.

Review questions are presented at the end of each chapter for practice. Exercises are included for readers to try their hands on the concepts and reinforce learning by doing. Case studies are presented to explain the practical application of the subjects covered, where possible. The chapters are organized in such a way that they are easy to reach, understand, and apply. We have given special emphasis to application instead of derivation of equations.

It must be mentioned that all pattern extraction and generation of mathematical equations have been performed using MS Excel. Statistical functions readily available have been used, and the use has been illustrated with solved examples. In some cases, we programmed the model equations in Excel.

All the equations used in this book have been tried out with software engineering data. These equations work. We have verified them and applied them to real-life problems.

We have taken utmost care to cite references and acknowledge contributions from experts. In case we have missed any, it is entirely due to oversight and we shall be obliged if such omissions are brought to our notice for correction.

We welcome feedback from readers which can be mailed to the email ids of the authors:

aravind_55@yahoo.com

skmurali7@yahoo.com

Authors

Ravindranath Pandian C. Quality Improvement Consultants (QIC) was founded in 1997 by Mr. C. Ravindranath Pandian. He has consulted to and trained many industries on innovation, project management, and quality management helping them to reap rich benefits.

Mr. Pandian has authored two books, *Software Metrics* and *Risk Management*, both published by CRC Press, USA.

Mr. Pandian has a keen interest in developing people and giving them the needed knowledge resources. He has always cherished training as a process. He views it as a social responsibility and a noble service. He has so far trained over 14,000 engineering and management professionals on quantitative methods in management. He teaches Six Sigma, project management, software metrics, and SPC. He visits colleges and teaches management graduates these value-adding concepts. He has solved numerous metrics data analysis problems in his workshops and has shown the way to thousands.

He speaks and conducts trainings at public forums such as SPIN (*Software Process Improvement Network Chapters* in Bangalore, Chennai, Hyderabad, Pune, and Trivandrum) and PMI (*Project Management Institute Chapters* such as in Singapore) to promote these values. His teachings have been acclaimed as "eye openers" by many. Software managers have appreciated the way he connects statistics with software development.

Mr. Pandian has held various positions such as GM-Corporate QA and Lead Assessor for ISO 9000 Quality Systems, and he has been engaged in R&D management in antennas, infrared pyrometers, telecommunications, and analytical instrumentation industry. He studied Electronics and Communication Engineering at MIT Chennai.

 **Murali Kumar S K**. Murali is well experienced in software quality, process, and product certifications of software/IT for aerospace, life sciences, and telecom industries.

He has multidisciplinary cross-functional experience in various industry environments like software/IT, electronics, metallurgical, mechanical, and electronics driving change management in Quality and Process.

He has worked with the Software Productivity Center of Honeywell on quality initiatives. He has deployed various quality models and standards like CMMI, PCMM, ISO 9001, and AS9100. As an Industry Experienced Lead Auditor in Aerospace, Murali has wide experience in auditing global aerospace supply chain sites operating in a multicultural global environment.

Murali is a certified Six Sigma Black Belt from Honeywell and mentored at least 50 Green Belt projects and trained more than 1000 people in Six Sigma. He has been working on implementing lean practices and continous improvement of Quality Systems.

Murali has a Bachelor's degree in Metallurgical Engineering from PSG College of Technology and MBA degree in Business Administration from Anna University. He is trained in General Management Program at Indian Institute of Management, Bangalore (IIM-B).

DATA

I

Data are where science and statistics begin. In software engineering, data-based decision making makes the difference between maturity and immaturity, professionalism and unprofessionalism. Data contain the seeds of knowledge. Data must be fostered and used. In Section I of this book, we present basic properties of data and discuss data quality.

In Chapter 1, we discuss data and descriptive statistics, a smart way to summarize data and see the hidden meaning. Chapter 2 is about detecting truth in data by spotting its central tendency. Chapter 3 presents ways of understanding data dispersion. Chapter 4 is devoted to Tukey's box plot, a brilliant exploratory data analysis tool. Data, once collected, must be processed by the techniques given in these four chapters.

Chapter 1

Data, Data Quality, and Descriptive Statistics

The Challenge That Persists

Data refer to facts in contrast to opinion or conjecture. Data are evidence, results, an expression of reality, and all such concrete realizations. Data are the result of observation and measurement (of life in general) of processes and products in software development. We use the term data to represent the basic measures (raw) and derived (manipulated) metrics.

Data collection remains a challenge even after fifty years of history. The challenge engulfs the two types of data collection: the routine data collection and the special purpose data collection, such as in improvement programs and experiments. Problems in these areas are more in the first kind. A summary of the problems in data collection was presented by Goethert and Siviy [1], who find that "inconsistent definitions of measures, meaningless charts" are among the reasons for poor data.

They complain that "the big picture is lost." We miss the forest for the trees. They also point fingers at a serious problem: the context of the indicators is not understood. Not finding a context for data is becoming a major crisis.

Data have no meaning apart from their context.

Shewhart

The routine data collection can be studied from five contexts, viewing from five management layers: business management, project management, process management, product management, and the recently introduced subprocess management. When data lose their context, they are considered irrelevant and are thus dismissed. When managers lose interest, data are buried at the source. The solution is to categorize metrics according to context and assure relevance to the stakeholders. The periodical metrics data report should be divided into interlinked and context-based sections. Different stakeholders read different sections of the report with interest. Context setting should be performed before the goal question metric (GQM) paradigm is applied to the metrics design.

Several software development organizations prefer to define "mandatory data" and call the rest as "optional data." Mandatory metrics are chosen from the context of the organization, whereas optional metrics are for local consumption. Mandatory metrics are like the metrics in a car dashboard display; the industry needs them to run the show. The industry chooses mandatory metrics to know the status of the project and to assess the situation, departing from the confines of GQM paradigm in response to operational requirements.

SEI's GQ(I)M framework [2] improved the GQM methodology in several ways. Using mental models and including charts as part of the measurement process are noteworthy. Instant data viewing using charts connects data with decision making and triggers biofeedback. Creating charts is a commendable achievement of statistical methods. Spreadsheets are available with tools to make adequate charts.

Mapping is frequently used in engineering measurements. The mapping phase of software size measurement in COSMIC Function Points is a brilliant exposition of this mapping. The International Function Point Users Group defines counting rules in a similar vein. Counting lines of code (LOC) is already a long established method. Unambiguous methods are available to measure complexity. These are all product data regarded as "optional." Despite the clarity provided by measurement technologies, product data are still not commonly available.

Moving up, business data include key performance indicators, best organized under the balanced scorecard scheme. These data are driven by strategy and vision and used in a small number of organizations as a complement to regular data collection.

Data availability has remained a problem and is still a problem. The degree of data availability problem varies according to the category of data. A summary is presented in the following table:

Category	Data Availability
1. Business data	Medium availability
2. Project data	High availability
3. Process data	Low availability
4. Subprocess data	Extremely low availability
5. Product data	Very low availability

Collecting data in the last two categories meets with maximum resistance from teams because this data collection is considered as micromanagement. The previously mentioned profile of data availability is typical of software business and contrasts with manufacturing; for example, product data are easily available there.

Bringing Data to the Table Requires Motivation

A strong sense of purpose and motivation is required to compile relevant data for perusal, study, and analysis. Different stakeholders see different sections of data as pertinent. Business managers would like to review performance data. Project managers would like to review delivery related data, including some performance data they are answerable to. Process managers would like to review process data. Model builders and problem solvers dig into subprocess data. An engineering team would be interested in looking at product data.

Data are viewed by different people playing different roles from different windows. Making data visible in each window is the challenge. The organizational purpose of data collection should be translated into data collection objectives for different stakeholders and different users. Plurality of usage breeds plurality in usage perspectives. Plurality is a postmodern phenomenon. Single-track data compiling initiatives fail to satisfy the multiple users, resulting in dismally poor metric usage across the organization.

The mechanics of data compilation and maintaining data in a database that would cater to diverse users is now feasible. One can look up data warehouse technology to know the method. A common, structured platform, however, seems to be a goal-driven process to bring data to the data warehouse.

Data Quality

On Scales

Software data have several sources as there are several contexts; these data come in different qualities. A very broad way of classifying data quality would be to divide data into qualitative and quantitative kinds. Verbal descriptions and subjective ratings are qualitative data. Numerical values are quantitative data. Stevens [3] developed scales for data while working on psychometrics, as follows: nominal, ordinal, interval, and ratio scales. The first two scales address qualitative data. The remaining two address quantitative data. Stevens restored legitimacy for qualitative data and identified permissible statistical analyses for each scale. Each scale is valuable in its own way, although most analysts prefer the higher scales because they carry data with better quality and transfer richer information.

When data quality is low we change the rules of analyses; we do not discard the data.

Steven's measurement theory has cast a permanent influence in statistical methods.

The lower scales with allegedly inferior data quality found several applications in market research and customer satisfaction (CSAT) measurement. CSAT data are collected in almost every software project, and an ordinal scale designed by Likert [4] is extensively used at present for this purpose. We can improve CSAT data quality by switching over to the ratio scale, as in the Net Promoter Score approach invented by Frederick [5] to measure CSAT. CSAT data quality is our own making. With better quality, CSAT data manifest better resolution that in turn supports a comprehensive and dependable analysis.

The advent of artificial intelligence has increased the scope of lower scale data. In these days of fuzzy logic, even text can be analyzed, fulfilling the vision of the German philosopher Frege, who strived to establish mathematical properties of text. Today, the lower scales have proved to be equally valuable in their ability to capture truth.

Error

All data contain measurement errors, whether the data are from a scientific laboratory or from a field survey. Errors are the least in a laboratory and the most in a field survey. We repeat the measurement of a product in an experiment, and we may get results that vary from trial to trial. This is the "repeatability" error. If many experimenters from different locations repeat the measurement, additional errors may appear because of person to person variation and environmental variation known as "reproducibility" error. These errors, collectively called *noise*, in experiments can be minimized by replication.

The discrepancy between the mean value of measured data and the true value denotes "bias." Bias due to measuring devices can be corrected by calibrating the devices. Bias in estimation can be reduced by adopting the wide band Delphi method. Bias in regularly collected data is difficult to correct by statistical methods.

Both bias and noise are present in all data; the magnitude varies. Special purpose data such as those collected in experiments and improvement programs have the least. Data regularly collected from processes and products have the most. If the collected data could be validated by team leaders or managers, most of the human errors could be reduced. Statistical cleaning of data is possible, to some extent, by using data mining approaches, as shown by Han and Kamber [6]. Hundreds of tools are available to clean data by using standard procedures such as auditing, parsing, standardization, record matching, and house holding. However, data validation by team leaders is far more effective than automated data mining technology. Even better is to analyze data and spot outliers and odd patterns and let these data anomalies be corrected by process owners. Simple forms of analysis such as line graphs, scatter plots, and box plots can help in spotting bad data.

Cleaned data can be kept in a separate database called a *data warehouse*. Using data warehouse techniques also help in collecting data from heterogeneous sources and providing data a structure that makes further analysis easy. The need for a commonly available database is felt strongly in the software industry. More and more data get locked into personal databases of team members. Although data collection is automated and data quality is free from bias and noise, the final situation is even worse: data are quietly logged into huge repositories with access available only to privileged managers. They do not have the time for data related work. The shoemaker syndrome seems to be working.

Data Stratification

This is one of the earliest known methods. Data must be grouped, categorized, or stratified before analysis. Data categories are decided from engineering and management standpoint. This should not be left to statistical routines such as clustering or principal component analysis.

In real life, stratification is performed neither with the right spirit nor with the required seriousness. For instance, a common situation that may be noticed is attempts to gather software productivity data and arriving at an organizational baseline. Productivity (function point/person month) depends on programming language. For example, Caper Jones [7] has published programming tables, indicating how the level of language increases as productivity increases.

Visual Summary

Descriptive statistics is used to describe and depict collected data in the form of charts and tables. Data are summarized to facilitate reasoning and analysis. The first depiction is the visual display of data, a part of indicators in the GQ(I)M paradigm [1]. The second depiction is a numerical summary of data.

Visual display is an effective way of presenting data. It is also called *statistical charting*. Graphical form communicates to the human brain better and faster, allowing the brain to do visual reasoning, a crucial process for engineers and managers. Park and Kim [8] proposed a model for visual reasoning in the creative design process. There is growing evidence to show that information visualization augments mental models in engineering design (Liu and Stasko [9]). Data visualization is emerging into a sophisticated discipline of its own merit.

Let us see as an example two simple graphs. First is a radar chart of project risks shown in Figure 1.1.

This provides a risk profile of project environment at a glance. The radar chart presents an integrated view of risk; it is also an elegant summary. This chart can be refreshed every month, showing project managers the reality. Reflecting upon the chart, managers can make decisions for action. The second chart is a line graph

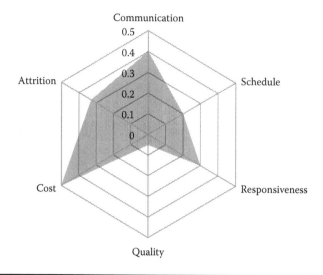

Figure 1.1   Radar chart for project risks.

of cumulative count of code written till date. The actual code written is plotted alongside the plan in Figure 1.2. By visual reasoning upon the plot, one can guess the time to finish the project.

Data must be transformed into charts, till then they do not enter decision space.

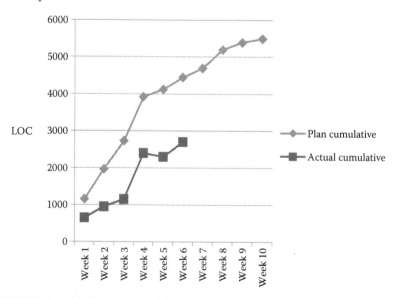

Figure 1.2   Cumulative count of code.

Even lower-scale data can be graphed. For example, a bar graph on discovered defect types can be very instructive. Most categorical variables are plotted as bar graphs and pie charts, and they make a lot of sense.

The graphs must be interpreted. A picture is worth a thousand words; but each one needs a few words of explanation articulating the context and meaning. Commentaries on graphs are rare; it may perhaps be assumed that truth is self-evident in the graphs. However, it makes a huge difference to add a line of comment to a graph.

BOX 1.1   SHOW ME A GRAPH

This organization was dedicated to software maintenance. Every month, a huge list of change requests are received. The operations manager found "backlog" a burning issue. The backlog seemed to grow every month. After due contemplation, he devised a simple management technique to address this issue. He suggested a simple pie chart report at the end of every month. The pie chart showed distribution of bugs according to the following category:

a. Bugs taken up—complex category
b. Bugs taken up—simple category
c. Bugs analyzed but found as nonissues
d. Bugs in queue—yet to be taken up
e. Bugs delivered

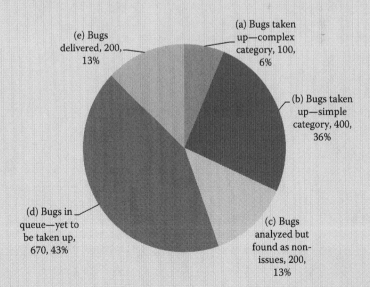

The pie chart had a noteworthy consequence. The backlog queue dwindled, and more bugs were fixed monthly. Later, the manager happened to know about "visual management" and ascribed success of the pie chart to visual management.

The pie chart was so simple and yet so effective; it soon became a weekly report and became very popular. The pie chart turned the company around.

Numerical Descriptive Statistics (Numerical Summary of Data)

The numerical summary of data has a standard set of statistics. There is a difference between data and statistic. Data are a result of measurement. Statistic is a result of statistical processing of data. There is a prerequisite for doing descriptive statistics. We need a set of observations—a sample of data points—to prepare a numerical summary. A few components should have been made or a few executions of a process should have been made before we think of a numerical summary. This constraint is not imposed on graphs. Data 1.1 presents the data sample and shows the effort variance data in a typical development project.

What does the data mean? Quantitative reasoning begins with a statistical inquiry into effort variance. What is the center of the process? What is the range of the process? Is the process data symmetrical as one would expect, or is it skewed? Does the process have a strong peak or is it flat? The answers to such queries are formally available in the form of some basic statistics. These statistics have been computed for the effort variance data using the Excel Data Analysis Tool "Descriptive Statistics." Data 1.2 presents the report from this tool.

Data 1.1 Effort Variance Data (%)

20.0
12.4
18.0
30.0
5.0
12.0
15.0
0.4
−3.0
4.0
7.0
9.0
10.0
6.0

Data 1.2 Descriptive Statistics of Effort Variance

Data	Descriptive Statistics	
20.0	Mean	10.41429
12.4	Standard error	2.278081
18.0	Median	9.5
30.0	Mode	N/A
5.0	Standard deviation	8.5238
12.0	Sample variance	72.65516
15.0	Kurtosis	0.908722
0.4	Skewness	0.720095
−3.0	Range	33
4.0	Minimum	−3
7.0	Maximum	30
9.0	Sum	145.8
10.0	Count	14
6.0	Confidence level (95.0%)	4.921496

There are fourteen basic "statistics" in the table. We can add the kth largest and kth smallest values to this list by ticking off the options in the tool. Definitions of these statistics are presented in Appendix 1.1.

BOX 1.2 POWER OF TABLE

Managing software development is a complex task. A manager applied data-driven management in a novel manner to make his task easy. He identified 12 milestones and selected the data he needed to collect for each milestone for effective management. That led him to design a data table with 12 rows and 10 columns. The data columns included dates, size defects, effort, and pertinent feature numbers. The milestones coincided with deliveries, and the data table came to be called the *milestone table*. With this simple table, he realized he could manage a project almost of any size and duration. He also found extra bandwidth to manage many more projects simultaneously. His teams never missed milestones because he took milestone level data seriously and reviewed the results objectively and with precision. His projects were often delivered on time, with quality and within budget.

Data Table		Project Name Customer Ref.							
1	2	3	4	5	6	7	8	9	10
Milestone	Delivery	Features	Start DT	Finish DT	Dev Effort	Review Effort	Test Effort	Test Defects	UAT Defects
1	Start	Architecture							
2	Package 1	F1–F5							
3	Package 2	F6–F20							
4	Package 3	F21–F40							
5	Package 4	F41–F50							
6	Package 5	F51–F67							
7	Package 6	F68–F73							
8	Package 7	F74–F85							
9	Package 8	F86–F91							
10	Package 9	F92–F100							
11	Package 10	F101–F104							
12	End	Integration							

Special Statistics

A few special statistics are explained in later chapters. Standard error is described in Chapter 13. Confidence interval is described in Chapter 21. Percentiles, quartiles, and interquartile range are explained in Chapter 4. We can assemble our preferred statistics into the descriptive statistics banner.

Three Categories of Descriptive Statistics

The simple and most commonly used descriptive statistics can be divided into three categories and analyzed more deeply:

Central tendency (discussed in Chapter 2)
Dispersion (discussed in Chapter 3)
Tukey's five-point summary (discussed in Chapter 4)

Such a deeper exploration might be viewed as part of exploratory data analysis.

Case Study: Interpretation of Effort Variance Descriptive Statistics

Let us look at the descriptive statistics of effort variance data provided in Data 1.2. The number of data points is 14. We would have preferred more than 30 data points for drawing conclusions. We can do with 14, keeping in mind that there could be small but tolerable errors in our judgment.

Two statistics are of significant consequence—the mean value is 10.414 and the maximum value is 30. We are going to apply business rules to evaluate these statistics and not statistical rules. The mean value of variance, when the estimation process is mature, should be close to zero. The ideal behavior of estimation errors is like that of measurement errors; both should be symmetrically distributed with the center at zero. After all, estimation is also a measurement. The current mean variance of 10.414 is high, suggesting that the project consistently loses approximately 10% of manpower. This is what Juran called *chronic waste*.

The second problem is that the maximum value of variance stretches as far as 30%. This is not terribly bad, from a practical angle. Projects have reported much higher extremities going once in a while as far as 80%. This is anyway a less serious problem than the mean value.

Both kurtosis and skewness are not alarming.
The median stays closer to the mean, as expected.
There is no clear mode in the data.

The range is 33, but the standard deviation is approximately 8.5, suggesting a mathematical process width of six times standard deviation, equal to 51. The mathematical model predicts larger variation of process. However, even this larger forecast is not alarming as the mean value.

Overall, the project team has a reasonable discipline in complying with plans, indicated by acceptable range. The estimation process requires improvement, and it looks as if the estimation process could be fine-tuned to achieve a mean error of zero.

BOX 1.3   SMALL IS BIG

The maintenance projects had to deal with 20,000 bugs every week pouring in from globally located customer service centers. The product was huge, and multiple updates happened every month and delivered to different users in different parts of the world. The maintenance engineers were busy fixing the bugs and had no inclination to look at and learn from maintenance data. The very thought of a database with millions of data points deterred them from taking a dip into the data. Managers were helpless in this regard because they had no case to persuade people to take large chunks of time and pore over data. Data were unpopular until people came to know about five-point summaries. A month's data can be reduced to Tukey's five statistics: minimum, first quartile, median, third quartile, and maximum. People found it very easy at merely five statistics to understand a month's performance.

Time to Repair Analysis
Tukey's Five-Point Summary

N (Bugs Reported)	Zone 1	Zone 2	Zone 3	Zone 4
	22,000	12,000	23,600	32,000
Statistic	**Time to Repair, Days**	**Time to Repair, Days**	**Time to Repair, Days**	**Time to Repair, Days**
Minimum	12	6	45	25
Quartile 1	70	44	57	63
Median	120	66	130	89
Quartile 3	190	95	154	165
Maximum	300	126	200	223

Application Notes

A primary application of the ideas we have seen in this chapter is in presenting data summaries. The design of summary tables deserves attention.

First, presenting too many metrics in a single table must be avoided. Beyond seven metrics, the brain cannot process parallel data. Data summary tables with 40 metrics go overhead. Such data can be grouped under the five categories: business, project, process, subprocess, and product. If such a categorization is not favored, the summary table can have any of the following categories:

Long term–short term
Business–process
Project–process
Project–process–product

What is important is that the table must be portioned into tiles; the parts may be presented separately connected by digital links. This way, different stakeholders may read different tables. Whoever picks up a table will find the data relevant and hence interesting.

Next, for every metric, the five-point summary may be presented instead of the usual mean and sigma for one good reason: most engineering data are nonnormal. The five-point summary is robust and can handle both normal and nonnormal data.

Concluding Remarks

It is important to realize the context of data to make both data collection and interpretation effective enough.

Before analyzing data, we must determine its scale. Permissible statistical methods change with scale. For example, we use median and percentiles for ordinal data.

Errors in data undermine our confidence in the data. We should not unwittingly repose undue confidence in data. We must seek to find the data sources and make an assessment of possible percentage of error in data. For example, customer perception data are likely to be inconsistent and subjective. In this case, we would trust the central tendency expressions rather than dispersion figures. Machine-collected data such as bug repair time is likely to be accurate.

We should learn to summarize data visually as well as numerically. We can make use of Excel graphs for the former and descriptive statistics in Excel for the latter. These summaries also constitute first-level rudimentary analyses without which data collection is incomplete.

Data have the power to change paradigms. Old paradigms that do not fit fresh data are replaced by new paradigms that fit. Data have the power to renew business management continually. Data are also a fertile ground for innovation, new discoveries, and improvement. All these advantages can be gained with rudimentary analyses of data.

BOX 1.4 ANALOGY: BIOFEEDBACK

There was this boy who stammered and went to a speech therapist. The treatment given was simple: he had to watch his speech waveform in an oscilloscope as he was speaking to a microphone. He practiced for 5 days, half an hour a day, and walked away cured of stammering. The way he gained normal speech is ascribed to biofeedback. Human systems correct themselves if they happen to see their performance. That is precisely what data usage in software development project achieves. When programmers see data about their code defects, the human instinctive capability is to rectify the problems and offer defect-free code. This principle has universal application and is relevant to all software processes, from requirement gathering to testing.

Review Questions

1. What are data?
2. What are scales of measurement?
3. What is a statistic? How is it different from data?
4. What are the most commonly used descriptive statistics?
5. What is Tukey's five-point summary?
6. How does data contribute to self-improvement?

Exercises

1. If you are engaged in writing code for a mission critical software application, and if you wish to control the quality of the code to ensure delivery of defect free components, what data will you collect? Design a data collection table.
2. During testing of a 5000 LOC code, what data will you collect for the purpose of assessing code stability?

Appendix 1.1: Definition of Descriptive Statistics

Number of Data Points

When we see a metric value, we should also know the size of the sample used in the calculation.

Number of data points (observations) n

Sum

This is a plain total of all values, useful as a meta-calculation:

$$\text{Sum} = \sum_{i=1}^{i=n} x_i$$

Variance

This is a mathematical calculation of data dispersion obtained from the following formula:

$$\text{Variance} = \frac{\sum_{i=1}^{n} (x_i - \bar{x})^2}{n-1}$$

where n is the sample size and $\bar{x}$ is the sample mean. Variance is the average squared deviation from the mean.

Standard Deviation

Square root of variance is equal to standard deviation. This is the mathematical expression of dispersion. This is also a parameter to normal distribution.

The standard deviation symbol σ is used to show the standard deviation notation. Symbol = σ, σ read as sigma:

$$\sigma = \sqrt{\text{variance}}$$

Maximum

This is the largest value in the sample. Large values of effort variance indicate a special problem and are worth scrutiny. The questions here are "How bad is the worst value? Is it beyond practical limits?" This statistic is a simple recognition of a serious characteristic of data.

Minimum

This is the other end of data values. The question is similar: "How low is the minimum value?" In effort variance, the minimum value can have a negative sign, suggesting cost compression. Usually, cost compression is good news, but process managers get cautious when the value becomes deeply negative. The questions that bother them are as follows: Has there been some compromise? Will cost saving have a boomerang effect?

Range

Range is obtained by subtracting the minimum from the maximum. Range represents process variation, in an empirical sense. This statistic is widely used in process control. It is simple to compute and yet sensitive enough to alert if processes vary too much.

Range is just the difference between the largest and the smallest values:

$$\text{Range} = \text{maximum} - \text{minimum}$$

Mode

Mode is the most often repeated value. It is an expression of central tendency.

Median

Median is the value that divides data—organized into an ordered array—into two equal halves. This is another expression of central tendency.

In simple words, median is the middle value in the list of numbers. A list should be arranged in an ascending order first to calculate the median value. Then the formula is stated as follows:

If the total number of numbers (n) is an odd number, then the formula is given as follows

$$\text{Median} = \left(\frac{n+1}{2}\right)^{\text{th}} \text{term}$$

If the total number of the numbers (n) is an even number, then the formula is as follows:

$$\text{Median} = \frac{\left(\frac{n}{2}\right)^{th} \text{term} + \left(\frac{n}{2}+1\right)^{th} \text{term}}{2}$$

Mean

Mean is the arithmetic average of all data points. This is an expression of central tendency. This is also a parameter to normal distribution:

$$\bar{x} = \frac{\sum x}{n}$$

Kurtosis (Flatness of Distribution)

Kurtosis is how peaked the data distribution is. Positive kurtosis indicates a relatively peaked distribution. Negative kurtosis indicates a relatively flat distribution (see Chapter 3 for the formula).

Skewness (Skew of Distribution)

Skewness is a measure of asymmetry in data. Positive skewness indicates a distribution with an asymmetric tail extending toward more positive values. Negative skewness indicates a distribution with an asymmetric tail extending toward more negative values (see Chapter 3 for the formula).

 References

1. W. Goethert and J. Siviy, *Applications of the Indicator Template for Measurement and Analysis*, SEI Technical Note CMU/SEI-2004-TN-024, 2004.
2. R. E. Park, W. B. Goethert and W. A. Florac, *Goal Driven Software Measurement—A Guidebook*, SEI Handbook CMU/SEI-96-HB-002, 1996.
3. S. S. Stevens, On the theory of scales of measurement, *Science*, 103, 677–680, 1946.
4. R. Likert, A technique for the measurement of attitudes, *Archives of Psychology*, 140, 1932.
5. F. F. Reichheld, *The One Number You Need To Grow*, Harvard Business Review, December 2003.
6. J. Han and M. Kamber, *Data Mining—Concepts and Techniques*, Morgan Kaupmann Publishers, 2nd Edition, 2006.

7. C. Jones, *Programming Productivity*, McGraw-Hill Series, New York, 1986.
8. J. A. Park and Y. S. Kim, Visual reasoning and design processes, *International Conference on Engineering Design*, 2007.
9. Z. Liu and T. J. Stasko, Mental models, visual reasoning and interaction in information visualization: A top-down perspective, *IEEE Transactions on Visualization and Computer Graphics*, 16, 999–1008, 2010.

Suggested Readings

Aczel, A. D. and J. Sounderpandian, *Complete Business Statistic*, McGraw-Hill, London, 2008.

Crewson, P., *Applied Statistics Handbook*, Version 1.2, AcaStat Software, 2006.

Downey, A. B., *Think Stats Probability and Statistics for Programmers*, Version 1.6.0, Green Tea Press, Needham, MA, 2011.

Dyba, T., V. B. Kampenes and D. I. K. Sjøberg, A systematic review of statistical power in software, *Information and Software Technology*, 48, 745–755, 2006.

Gupta, M. K., A. M. Gun and B. Dasgupta, *Fundamentals of Statistics*, World Press Pvt. Ltd., Kolkata, 2008.

Hellerstein, J. M., *Quantitative Data Cleaning for Large Databases*, EECS Computer Science Division, UC Berkeley, United Nations Economic Commission for Europe (UNECE), February 27, 2008. Available at http://db.cs.berkeley.edu/jmh.

Holcomb, Z. C., *Fundamentals of Descriptive Statistics*, Pyrczak Publishing, 1998.

Lussier, R. N., *Basic Descriptive Statistics for Decision Making*, e-document.

NIST/SEMATECH, *Engineering Statistics Handbook*, 2003. Available at http://www.itl.nist.gov/div898/handbook/.

Shore, J. H., *Basic Statistics for Trainers*, American Society for Training & Development, Alexandria, VA, 2009. Available at http://my.safaribooksonline.com/book/statistics/9781562865986.

Succi, G., M. Stefanovic and W. Pedrycz, *Advanced Statistical Models for Software Data*, Department of Electrical and Computer Engineering, University of Alberta, Edmonton, AB, Canada. Proceedings of the 5th World Multi-Conference on Systemics, Cybernetics and Informatics, Orlando, FL, 2001. Available at http://www.inf.unibz.it/~gsucci/publications/images/advancedstatisticalmodelsforsoftwaredata.pdf.

Tebbs, J. M., *STAT 110 Introduction to Descriptive Statistics*, Department of Statistics, University of South Carolina, 2006. Available at http://www.stat.sc.edu/~tebbs/stat110/fall06notes.pdf.

Torres-Reyna, O., *Data Preparation & Descriptive Statistics*, Data Consultant. Available at http://www.princeton.edu/~otorres/DataPrep101.pdf.

Chapter 2

Truth and Central Tendency

We have seen three statistical expressions for central tendency: mean, median, and mode. *Mean* is the arithmetic average of all observations. Each data point contributes to the mean. *Median* is the middle value of the data array when data are arranged in an order—either increasing order or decreasing order. It is the value of a middle position of the ordered array and does not enjoy contribution from all observations as the mean does. *Mode* is the most often repeated value. The three are equal for symmetrical distributions such as the normal distribution. In fact, equality of the three values can be used to test if the data are skewed or not. Skew is proportional to the difference between mean and mode.

Mean

Use of mean as the central tendency of data is most common. The mean is the true value while making repeated measurements of an entity. The way to obtain truth is to repeat the observation several times and take the mean value. The influence of random errors in the observations cancel out, and the true value appears as the mean. The central tendency mean is used in normal distribution to represent data, even if it was an approximation. Mean is the basis for normal distribution; it is one of the two parameters of normal distribution (the other parameter is standard deviation). One would expect the mean value of project variance data such as effort variance, schedule variance, and size variance to reveal the true error in estimation.

Once the true error is found out, the estimation can be calibrated as a measurement process.

It is customary to take a sample data and consider the mean of the sample as the true observation. It makes no statistical sense to judge based on a single observation. We need to think with "sample mean" and not with stray single points. "Sample mean" is more reliable than any individual observation. "Sample mean" dominates statistical analysis.

Uncertainty in Mean: Standard Error

The term "sample mean" must be seen with more care; it simply refers to the mean of observed data. Say we collect data about effort variance from several releases in a development project. These data form a sample from which we can compute the mean effort variance in the project. Individual effort variance data are used to measure and control events; sample mean is used to measure and control central capability. Central tendency is used to judge process capability.

Now the Software Engineering Process Group (SEPG) would be interested in estimating process capability from an organizational perspective. They can collect sample means from several projects and construct a grand mean. We can call the grand mean by another term, the population mean. Here population refers to the collective experience of all projects in the organization. The population mean represents the true capability of organization.

If we go back to the usage of the term truth, we find there are several discoveries of truth; each project discovers effort variance using sample mean. The organization discovers truth from population mean.

Now we can estimate the population mean (the central tendency of the organizational process) from the sample mean from one project (the central tendency of the local process). We cannot pinpoint the population mean, but we can fix a band of values where population mean may reside. There is an uncertainty associated with this estimation. It is customary to define this uncertainty by a statistic called *standard error*. Let us look further into this concept.

It is known that the mean values gathered from different projects—the sample means—vary according to the normal distribution. The theorem that propounds this is known as the *central limit theorem*. The standard deviation of this normal distribution is known as the *standard error*.

If we have just collected sample data from one project with n data points, and with a standard deviation s, then we can estimate standard error with reasonable accuracy using the relation

$$SE = \frac{s}{\sqrt{n}}$$

Defining an uncertainty interval for mean is further explained in Chapter 25.

Median

The physical median divides a highway into two, and the statistical median divides data into two halves. One half of the data have values greater than the median. The other half of the data have values smaller than the median. It is a rule of thumb that if data are nonnormal, use median as the central tendency. If data are normally distributed, median is equal to mean in any case. Hence, median is a robust expression of the central tendency, true for all kinds of data. For example, customer satisfaction data—known as *CSAT data*—are usually obtained in an ordinal scale known as the *Likert scale*. One should not take the mean value of CSAT data; median is the right choice. (It is a commonly made mistake to take the mean of CSAT data.) In fact, only median is a relevant expression of central tendency for all subjective data. Median is a truer expression of central tendency than mean in engineering data, such as data obtained from measurements of software complexity, productivity, and defect density.

> While the mean is used in the design of normal distribution, the median is used in the design of skewed distributions such as the Weibull distribution. Median value is used to develop the scale parameter that controls width.

BOX 2.1   HANGING A BEAM

Think of mean as a center of gravity. In Figure 2.1, the center of gravity coincides with the geometric center, which is analogous to the median of the beam, and as a result, the beam achieves equilibrium. In Figure 2.2, the center of gravity shifts because of asymmetrical load distribution; the beam tilts in the direction of center of gravity. The median, however, is still the same old point. The distance between median and center of gravity is like the difference between median and mean. Such a difference makes the beam tilt; in the case of a data array, the difference between median and mean is a signal of data "skew" or asymmetry.

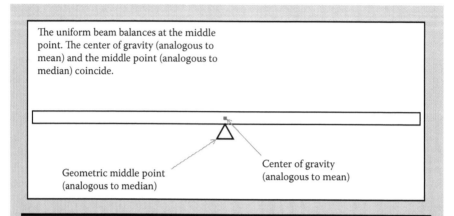

The uniform beam balances at the middle point. The center of gravity (analogous to mean) and the middle point (analogous to median) coincide.

Geometric middle point (analogous to median)

Center of gravity (analogous to mean)

Figure 2.1 Geometric middle point and center of gravity coincides and the beam is balanced.

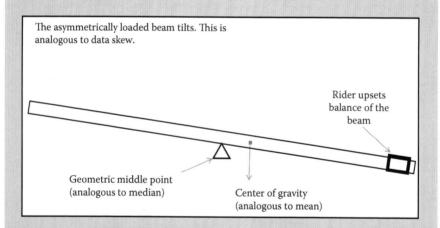

The asymmetrically loaded beam tilts. This is analogous to data skew.

Rider upsets balance of the beam

Geometric middle point (analogous to median)

Center of gravity (analogous to mean)

Figure 2.2 Asymmetry is introduced by additional weight on the rightside of the beam. The mean shifts to the right.

BOX 2.2   A ROBUST REFERENCE

Median is a robust reference that can serve as a baseline much better than mean serves. If we wish to monitor a process, say test effectiveness, first we need to establish a baseline value that is fair. Median value is a fair central line of the process, although many tend to use mean. Mean is already influenced by extreme values and is "prejudiced." Median reflects true performance of the process. Untrimmed mean reflects the exact location of the process without any discrimination. Median effectively filters away prejudices and offers a fair and robust judgment of process tendency. For example, the median score of a class in a given subject is the true performance of the class, and the mean score does not reflect the true performance.

Mode

Mode, the most often repeated value in data, appears as the peak in the data distribution. Certain judgments are best made with mode. The arrival time of an employee varies, and the arrival data are skewed as indicated in the three expressions of central tendency: mean = 10:00 a.m., confidence interval of the mean = 10:00 a.m. ± 20 minutes, median = 9:30 a.m., and mode = 9:00 a.m. The expected arrival time is 9:00 a.m. Let us answer the question, is the employee on time? The question presumes that we have already decided not to bother with individual arrival data but wish to respond to the central tendency. Extreme values are not counted in the judgment. We choose the mode for some good reasons. Mean is biased by extremely late arrivals. Median is insensitive to best performances. Mode is more appropriate in this case.

Geometric Mean

When the data are positive, as is the case with bug repair time, we have a more rigorous way of avoiding the influence of extreme values. We can use the concept of geometric mean.

The geometric mean of n numbers is the nth root of the product of the n numbers, that is,

$$GM = \sqrt[n]{x_1 x_2 \ldots x_n}$$

Geometric mean can also be calculated from the arithmetic mean of the logarithm of the *n* numbers. Then this must be converted back to a "base 10 number" by using an antilogarithm.

A geometric mean, unlike an arithmetic mean, tends to mitigate the effect of outliers, which might bias the mean if a straight average (arithmetic mean) was calculated.

The geometric mean for bug repair time given in Table 2.1 is found to be 17.9. We can use the Excel function GEOMEAN to calculate this. In this case, it may be noted that the geometric mean is almost equal to the median value. It may be

Table 2.1   Bug Repair Time

Number of Days		
16	31	7
23	19	28
45	18	29
20	18	12
13	21	49
13	39	20
58	14	21
9	11	49
7	11	14
29	9	15
13	25	13
12	25	6
32	20	28
31	17	21
31	13	23
33	13	13
6	13	16
31	24	10
26	12	14
21	7	14

remembered that all data values are not used in the median computation, whereas every data value is used in the geometric mean.

There are certain financial return calculations where geometric mean is the right choice. If an investment earns 20% in the first year, 40% in the second year, 50% in the third year, and 60% in the fourth year, the average return is not the arithmetic mean of 42.5% but the geometric mean of 41.703%. It is an error to use the arithmetic mean in this case.

Jeff and Lewis [1] have studied tasks times in usability tests that are positively skewed. They report that the median does not use all the information available in a sample. Using the geometric mean, they have achieved 13% less error and 22% less bias than the median.

Harmonic Mean

With positive data, we have yet another statistic to yield central tendency without bias from extreme values: the harmonic mean. It is even more protective than geometric mean, that is,

$$HM = \frac{N}{\dfrac{1}{x_1} + \dfrac{1}{x_2} + ... + \dfrac{1}{x_N}}$$

To find the harmonic mean of a set of *n* numbers, we add the reciprocals of the numbers in the set, divide the sum by *n*, then take the reciprocal of the result. The harmonic mean is the reciprocal of the arithmetic mean of reciprocals. This gives further screening from extreme values. The harmonic mean for bug repair time data given in Table 2.1 is 15.6 days. This value is closer to the mode than the median, the geometric mean, or the mean.

The Excel function to find harmonic mean is HARMEAN.

A formal treatment of geometric and harmonic means may be found in the *Handbook of Means and Their Inequalities* by Bullen [2].

Interconnected Estimates

In interpreting the central tendency of software data, so much depends on the situation. In most cases, data are skewed; therefore, mean, median, and mode are different. In such cases, there is no one word answer to central tendency. There are three values that need to be studied and interpreted.

Consider the case of repair time of a particular category of bugs in a software development project. Bug repair time data are given in Table 2.1.

The following are the five values of central tendency:

Arithmetic mean	20.517
Median	18.000
Mode	13.000
Geometric mean	17.867
Harmonic mean	15.561

The team leader wants to set a goal for bug repair time and also wants to plan resources for the next quarter based on the bug arrival forecast. He wants to take data-driven decisions. He wants optimum decisions too. Which expression of truth will he use?

If we subscribe to the approach that people should follow best practices, the mode should be used to define goal. Aggressive goal setting can still be based on the best performance demonstrated: mode. We need a realistic value to be used in resource planning. We can either choose the median or the mean. Mean is safer and can provide a comfortable cushion. However, then we will be overplanning the resources. A look at the data set shows that maximum value is 58 days. We realize that such extreme values have biased mean values and deteriorated its application potential. Thus, the mean is rejected. A fair answer could be the median.

If the data are positive but skewed, then the geometric and harmonic means can be used. Hence, if the data are complex, we need to look at the multiple estimates of central tendency instead of just the mean.

Weighted Mean

There are times when we weight data *x* with factors *w* and find the weighted average using the following formula:

$$\bar{x} = \frac{\sum\limits_{i=1}^{n} w_i x_i}{\sum\limits_{i=1}^{n} w_i}$$

In the Program Evaluation and Review Technique (PERT) calculation, the estimated schedule is a weighted mean of three values:

Optimistic value	{O}	Weight 1
Pessimistic value	{P}	Weight 1
Most likely value	{ML}	Weight 4

Table 2.2 Expert Judgment of Milestone Schedule (Days)

	Weight	*Data*	*Weighted Data*
Optimistic	1	20	20
Most likely	4	40	160
Pessimistic	1	90	90
Average		50	45

$$\text{Estimate} = \frac{1O + 4ML + 1P}{1 + 4 + 1}$$

Expert judgment of a milestone schedule (days) is shown in Table 2.2. A proper estimate is obtained by applying weighted average.

In the previous example, the arithmetic mean is 50 days, and the weighted mean is 45 days.

Robust Means

The robust estimate of the mean is less affected by extreme values.

Trimmed Mean

Arithmetic mean breaks down if an extreme value is introduced. Even the presence of one extreme value can change this mean. In other words, it has a 0 breakdown point.

Trimming data gives us robust estimates of the mean, in the sense that the mean is resistant to changes in outlier data. Calculating the arithmetic mean after removing x% of data in the lower side and x% of data in the higher side will lead us to x% trimmed mean. Practically, x% can vary from 3% to 25%; x% is also called *breakdown point*.

In schools, the mean score of a class is calculated after removing 5% from the top and 5% from the bottom scores. It is 5% trimmed mean.

In process management, trimming is not a very straightforward step. Trouble in the process is normally revealed in the outliers. We identify outliers and do root cause analysis on them for process improvement. We cannot mindlessly discard extreme values while data cleaning. We can trim data to find a robust expression for central tendency, but the removed data have meaning elsewhere and need to be stored.

For more on trimmed means, refer to the thesis by Wu [3].

Winsorized Mean

Winsorized mean is similar to the trimmed mean. However, instead of trimming the points, they are set to the lowest or highest value. The beneficial properties of Winsorized means for skewed distributions are discussed by Rivest [4].

Midhinge

This is the average of the first and the third quartiles (the 25th and the 75th percentiles). This is a robust estimate.

Midrange

Midrange is the average of the smallest and the largest data.

Tukey's Trimean

This is obtained from the quartiles using the formula

$$\text{Trimean} = \frac{Q_1 + 2Q_2 + Q_3}{4}$$

Mean Derived from Geometrical Perspectives

Interesting geometric-based definitions of mean are summarized by Umberger [5]. Different means are seen as geometric properties of trapezoids.

Two Categories

We can divide expressions of central tendency into two categories. In the first category, we obtain participation from all observations in calculating central tendency. There are just three expressions that belong to this category. These measures naturally support mathematical modeling.

Category 1

1. Mean: we can use mean as a first-order judgment of central tendency
   Mean gives true value if we replicated an experiment.
   Estimating mean removes random noise.
   Mean provides a basis for building normal distribution from data.
   Mean is affected by extreme values.

2. Geometric mean: central tendency for skewed positive data
3. Harmonic mean: central tendency for skewed positive data

In the second category, we obtain participation from only strategically selected data points. We have seen seven such measures.

Category 2

1. Mode: a better indicator of central tendency in human performance
2. Median: a better indicator of central tendency in nonnormal data
3. Trimmed mean: straight removal of extreme values
4. Winsorized mean: robust calculation
5. Trimean: a weighted average of quartiles
6. Midhinge: an average of the first and the third quartiles
7. Midrange: average of lowest and largest values

We can estimate a pertinent set of means before judging the central tendency. The choice would depend on the skew, the presence of outliers, and the degree of protection we need from outliers. Such a choice would make analysis robust and safe.

Truth

Truth, expressed as central tendency, has many variants. We can narrow down our options depending on the type of data and depending on what we wish to do with the finding. The message is not in the mean, nor in the median. The message is to be seen in the many expressions. To the mathematically inclined, geometric mean and harmonic mean are alternatives to the arithmetic mean. The differences must be reconciled with practical reasoning.

> *Statistical judgment is never the ultimate end.*
> *Further reasoning alone can discover truth.*

Statistical calculations need not be the ultimate truth. At best, they can guide us toward truth. The moment of truth occurs only with reasoning.

To the empirical researchers, there is a series of trimmed means to bestow alternatives to the median.

The impact of multiple definitions of central tendency is rather heavy while evaluating shifts in process means. It is safer to work out all the definitions, obtain multiple numbers for central tendency, and treat them as a small universe of values.

> *We will have to compare one universe of central tendency values with another. We can no longer pitch one mean against another or engage in such misleading exercises.*

BOX 2.3 A GOLDEN RULE TO OBTAIN TRUTH

Estimating a software project using the "expert judgment" method is knowledge driven. However, selective memory could taint human judgment because knowledge is embedded in the human mind. A golden rule to extract truth from expert judgment is to make the expert recall extreme values as well as the central value from previous experience and use a weighted average using the 1:4:1 ratio. This estimate is respected as a golden estimate, and the rule is hailed as the golden rule.

Application Notes

Managing Software Projects Using Central Tendency Values

After collecting all the data, software projects are more commonly managed with values of central tendency. Managers prefer to take decision with summary truths.

Goal tracking is done using mean values while risk management is done using variances.

Weekly and monthly reports make liberal use of mean values of data collected. Performance dashboards make wide use of mean values. Means are compared to evaluate performance changes.

Making Predictions

Basic forecasts address mean values. Most prediction models present mean values. In forecasting business volumes and resource requirements, central tendencies are predicted and used as a rule. The prediction of variance is performed as a special case to estimate certainty and risk.

BOX 2.4 ALIGNING THE MEAN

Aligning the mean of results with target is a great capability. Process alignment with target is measured by the distance between mean and target. The lesser the distance, the greater is the alignment. Aligned processes synchronize with goals, harmonize work flow, and multiply benefits. The mean of results is particularly important in this context. The quality guru Taguchi mentions that the loss to society is proportional to the square of the drift of process mean from target, that is,

$$Loss = (target - mean)^2$$

Drift favorable to the consumer creates loss to the supplier; drift in the opposite direction creates loss to the consumer. Either way, drift causes loss to someone in society.

Case Study: Shifting the Mean

Performance is often measured by the mean. This is true even in the case of engineering performance. Code complexity is an engineering challenge. Left to themselves, programmers tend to write complex codes, a phenomenon known as *software entropy.* Code complexity is measured by the McCabe number. Shifting the mean value of the McCabe number in code requires drive from leaders and motivation from programmers. To make a shift in the mean complexity is a breakthrough in software engineering. Lower complexity results in modules that are testable and reliable. Nevertheless, achieving lower complexity requires innovation in software structure design and programming approaches.

This case study is about a software development project that faced this challenge and overcame it by setting a visible goal on mean complexity. The current state is defined by the mean McCabe number, and the goal state is defined by the desired McCabe number. The testing team suggested an upper limit of 70, beyond which code becomes difficult to comprehend in terms of test paths; test coverage also suffers. Data for the current state show huge variation in complexity from 60 to 123 and even 150 occasionally. The project manager has two thoughts: fix an upper limit on complexity of individual objects or fix an upper limit for the mean complexity number for a set of objects that define a delivery package. Although these two options look similar, the practical implications are hugely different. On the first option of setting an upper limit on individual events, the limit contemplated by testers is 70. The second option is about setting limit on the central tendency; this really is setting an optimum target for software development. The number chosen for this target is 40. This is a stretch goal, making the intention of the project manager very clear. Figure 2.3 shows the chart used by the project team to deploy this stretch goal for shifting the mean.

In reality, the team is gradually moving toward the targeted mean value. The direction of shift in the mean is very satisfying.

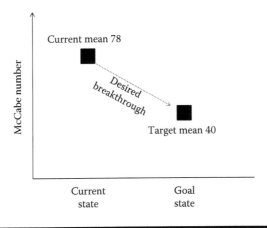

Figure 2.3   Reducing the code complexity.

Review Questions

1. What are the strengths and weaknesses of arithmetic mean?
2. What is the most significant purpose of using trimmed means?
3. Why is median considered more robust?
4. How will you find central tendency in ordinal data?
5. What are the uses of weighted average?

Exercises

1. Customer Satisfaction Data in a software development project is obtained in a Likert scale ranging from 1 to 5. The 1-year customer satisfaction scores are given below. Find the central tendency in the data (data 2, 4, 3, 5, 1, 3, 3, 4, 5, 3, 2, 1).
2. A test project duration is estimated by an expert who has made the following judgments:
     Optimistic duration: 45 days
     Pessimistic duration: 65 days
     Most likely duration: 50 days
   What do you think is the final and fair estimate of the test project duration?

References

1. J. Sauro and J. Lewis, Average task times in usability tests: What to report, *CHI 2010*, Atlanta, April 2010.
2. P. S. Bullen, *Handbook of Means and Their Inequalities*, Kluwer Academics Publisher, The Netherlands, 2003.
3. M. Wu, *Trimmed and Winsorized Estimators*, Michigan State University. Probability and Statistics Department, p. 194, 2006.
4. L. P. Rivest, Statistical properties of Winsorized means for skewed distributions, *Biometrika*, vol. 81 no. 2. pages no. 373–383, 1994.
5. S. Umberger, *Some Mean Trapezoids*, Department of Mathematics Education, University of Georgia, 2001.

Suggested Reading

Aczel, A. D. and J. Sounderpandian, *Complete Business Statistics*, McGraw-Hill, London, 2008.

Chapter 3

Data Dispersion

Data dispersion arises because of sources of variation, including variations in measurements and results due to changes in the underlying process. Which one varies more, measurement or processes under measurement? We proceed in this chapter with the assumption that there is sufficient measurement capability behind the data; that means the measurement errors are very small compared with process variation. If this condition is met, dispersion in data will represent dispersion in the process under measurement.

Range-Based Empirical Representation

Dispersion or variation in process is viewed as uncertainty in the process outcome. To deal with uncertainty, we need to measure, express, and understand it. Range is a good old way of measuring dispersion. This has been used in the traditional X-Bar and R control charts, where R stands for range and X-Bar stands for sample mean. Range is the difference between maximum and minimum values in data.

There is another convention to leave out extreme values and the considered range. Typically, the values below the 3rd percentile and the values above the 97th percentile are disregarded. To understand this, we need to construct the data array and sort data in some order and then chop off the upper 3rd percentile and the lower 3rd percentile. (If the length of the array is L, the 3rd percentile point will rest at a point on the array at a distance of $0.03L$ from the origin. Similarly, the point of the 97th percentile will rest at a distance of $0.97L$ from the origin.) This range is the empirical difference between the 3rd and the 97th percentile values.

The two calculations previously mentioned are conservative. In a third approach, the interquartile range (IQR) is taken as the core variation of the process. IQR is the difference between Q3 and Q1. It may be noted that 50% of observations remain in IQR.

Data 3.1   Design Effort Data

% Design Effort
2.87
11.55
11.11
18.55
21.08
100.00
83.56
6.75
6.33
9.71
21.25
6.02
30.63
13.14
26.16
18.85
32.14
10.59

Here is a summary of the three empirical expressions of dispersion:

$$Range = maximum-minimum$$

$$Percentile\ range = 97th-3rd\ percentile$$

$$IQR = Q_3 - Q_1$$

Example 3.1: Studying Variation in Design Effort

Design effort data as a percentage of project effort have been collected. The data are shown in Data 3.1.

The dispersion analysis of the data using the above-mentioned formulas is shown as follows:

% Design Effort	
Max	100.00
Min	2.87
Range	97.13
3rd percentile	4.478
97th percentile	91.616
Range	87.139
1st quartile	9.932
2nd quartile	24.933
Range	15.001

The full range, 97.13, is uncomplicated. The other two ranges have been trimmed. The 3rd percentile range (obtained by cutting off 3% of data on either side) is 87.139. The IQR is 15.001. Choosing the trimming rules has an inherent trouble—it can tend to be arbitrary, unless we exercise caution. Trimming the range is a practical requirement because we do not want extreme values to misrepresent the process. The untrimmed range is so large that it is impractical. There could be data outliers; one would suspect wrong entry, or wrong computation of the percentage of design effort. It is obvious that the very high values, such as 100% design effort, are impractical. Because the data have not been validated by the team, we can cautiously trim and "clean" the data. The 3rd percentile range seems to be clean, but it still has recognized a high value of 91.616%, again an impractical value. Perhaps we can tighten the trimming rule, say a cutoff of 20% on either end of data. If we want such tight trimmings, we might as well use the IQR, which has trimmed off 25% of data on either end.

Example 3.2: Analyzing Effort Variance Data from Two Processes

Let us take a look at effort variance data from two types of projects, following two different types of estimation processes. The data are shown in Data 3.2.

A summary of range analysis is shown as follows: Estimation A and Estimation B

	Estimation A	Estimation B
Full range	86.482	63.640
Percentile range	69.466	45.181
IQR	19.995	8.075

The three ranges individually confirm that the second set of data from projects using a different estimation model has less dispersion.

In both of the previously mentioned examples, we have studied dispersion from three different angles. We have not looked into the messages derived from the extreme values. Extreme values are used elsewhere in risk management and hazard analysis.

Range calculations approach dispersion from the extreme values—the ends—of data. These calculations do not use the central tendencies. In fact, these are independent of central tendencies.

Next we are going to see expressions of dispersion that consider central tendency of data.

BOX 3.1 CROSSING A RIVER

There was a man who believed in averages. He had to cross a river but did not know how to swim. He obtained statistics about the depth and figured out that the average depth was just 4 feet. This information comforted him because he was 6 feet tall and thought he could cross the river. Midway in the river, he encountered a 9-foot-deep pit and never came out. This story is often cited to caution about averages. This story also reminds us that we should register the extreme values in data for survival.

Data 3.2   Dispersion in Effort Variance Data

Effort Variance – 1	Effort Variance – 2
51.41	0.07
23.20	9.00
−7.67	−7.67
0.19	0.19
8.31	0.79
4.32	4.32
22.58	22.58
8.23	8.23
28.57	28.57
7.24	7.24
−1.39	−1.39
−1.31	−1.31
−11.32	−11.32
1.27	1.27
6.51	6.51
0.00	0.00
48.15	6.67
1.38	1.38
−2.67	−2.67
23.02	23.02
−35.07	−35.07
4.67	4.67

Range	Effort Variance – 1	Effort Variance – 2
	86.48	63.64

Percentile Range	Effort Variance – 1	Effort Variance – 2
3rd Percentile	−20.11	−20.11
97th Percentile	49.36	25.07
Percentile Range	69.47	45.18

Inter Quartile Range (IQR)	Effort Variance – 1	Effort Variance – 2
Q1	−0.98	−0.98
Q3	19.01	7.09
IQR	19.99	8.07

Dispersion as Deviation from Center

The range of data is a fairly good measure of dispersion. However, if we look at the scatter of data around a center, we will obtain a better and more complete picture. To visualize this, let us look at hits from a field gun on an enemy target. The hits are scattered around the target. If the gun is biased, the hits may be scattered around a center slightly away from the target. In both the scenarios, the hits are scattered around a center, the mean value.

Dispersion is measured in terms of deviations from the mean. Data with larger dispersion show larger deviations from the mean. The following are different expressions making use of this basic idea.

Average Deviation

If we calculate the deviations of all data points—all the N hits—from the mean $\bar{x}$ and take the average, we will obtain a measure of scatter, or dispersion. The average deviation for N hits from the field gun is a fairly representative estimate of dispersion. Every deviation is measured in meters, and the average deviation is also in meters.

The formula is given as follows:

$$D_{\bar{x}} = \frac{\sum_{i=1}^{N}(x - \bar{x})}{N} \tag{3.1}$$

When we try the same formula to obtain the average deviation in bug repair times (data shown in Table 3.1), we encounter a problem. The average deviation can be seen as zero.

The deviation values have a direction; the value can be positive or negative. In the process of calculating the average deviation, the positive values have cancelled out the negative values.

Average Absolute Deviation

To avoid the sign problem, we can take absolute values of deviation, as shown in this modified formula. This is a workaround.

$$D_{\bar{x}} = \frac{\sum_{i=1}^{N}|x_i - \bar{x}|}{N} \tag{3.2}$$

Absolute deviations have been calculated and shown in Table 3.1. The average absolute deviation from the mean is 8.669 days. This number defines dispersion of bug repair time around the data mean of 20.517.

We can use this measure to compare scatter in 2-month bug repair data.

Median Absolute Deviation

The scale of scatter of data can be computed with respect to any center. Normally, we use mean as the center as we have conducted in computing the average absolute

Table 3.1 Average Deviation and Average Absolute Deviation of Bug Repair Time

Bug Repair Time (days)	Mean (days)	Deviation from Mean (days)	Absolute Deviation from Mean (days)
16	20.517	−4.517	4.517
23	20.517	2.483	2.483
45	20.517	24.483	24.483
20	20.517	−0.517	0.517
13	20.517	−7.517	7.517
13	20.517	−7.517	7.517
58	20.517	37.483	37.483
9	20.517	−11.517	11.517
7	20.517	−13.517	13.517
29	20.517	8.483	8.483
13	20.517	−7.517	7.517
12	20.517	−8.517	8.517
32	20.517	11.483	11.483
31	20.517	10.483	10.483
31	20.517	10.483	10.483
33	20.517	12.483	12.483
6	20.517	−14.517	14.517
31	20.517	10.483	10.483
26	20.517	5.483	5.483
21	20.517	0.483	0.483
31	20.517	10.483	10.483
19	20.517	−1.517	1.517
18	20.517	−2.517	2.517
18	20.517	−2.517	2.517
21	20.517	0.483	0.483
39	20.517	18.483	18.483
14	20.517	−6.517	6.517
11	20.517	−9.517	9.517
11	20.517	−9.517	9.517
9	20.517	−11.517	11.517
25	20.517	4.483	4.483

(Continued)

Table 3.1 (Continued)   Average Deviation and Average Absolute Deviation of Bug Repair Time

Bug Repair Time (days)	Mean (days)	Deviation from Mean (days)	Absolute Deviation from Mean (days)
25	20.517	4.483	4.483
20	20.517	−0.517	0.517
17	20.517	−3.517	3.517
13	20.517	−7.517	7.517
13	20.517	−7.517	7.517
13	20.517	−7.517	7.517
24	20.517	3.483	3.483
12	20.517	−8.517	8.517
7	20.517	−13.517	13.517
7	20.517	−13.517	13.517
28	20.517	7.483	7.483
29	20.517	8.483	8.483
12	20.517	−8.517	8.517
49	20.517	28.483	28.483
20	20.517	−0.517	0.517
21	20.517	0.483	0.483
49	20.517	28.483	28.483
14	20.517	−6.517	6.517
15	20.517	−5.517	5.517
13	20.517	−7.517	7.517
6	20.517	−14.517	14.517
28	20.517	7.483	7.483
21	20.517	0.483	0.483
23	20.517	2.483	2.483
13	20.517	−7.517	7.517
16	20.517	−4.517	4.517
10	20.517	−10.517	10.517
14	20.517	−6.517	6.517
14	20.517	−6.517	6.517
Average Deviation		0.000	
Average Absolute Deviation			8.669

Data 3.3   Absolute Deviation of Bug Repair Time from the Median

Bug Repair Time Days	Median Days	Absolute Deviation Days	Bug Repair Time Days	Median Days	Absolute Deviation Days
16	18.000	2.000	25	18.000	7.000
23	18.000	5.000	20	18.000	2.000
45	18.000	27.000	17	18.000	1.000
20	18.000	2.000	13	18.000	5.000
13	18.000	5.000	13	18.000	5.000
13	18.000	5.000	13	18.000	5.000
58	18.000	40.000	24	18.000	6.000
9	18.000	9.000	12	18.000	6.000
7	18.000	11.000	7	18.000	11.000
29	18.000	11.000	7	18.000	11.000
13	18.000	5.000	28	18.000	10.000
12	18.000	6.000	29	18.000	11.000
32	18.000	14.000	12	18.000	6.000
31	18.000	13.000	49	18.000	31.000
31	18.000	13.000	20	18.000	2.000
33	18.000	15.000	21	18.000	3.000
6	18.000	12.000	49	18.000	31.000
31	18.000	13.000	14	18.000	4.000
26	18.000	8.000	15	18.000	3.000
21	18.000	3.000	13	18.000	5.000
31	18.000	13.000	6	18.000	12.000
19	18.000	1.000	28	18.000	10.000
18	18.000	0.000	21	18.000	3.000
18	18.000	0.000	23	18.000	5.000
21	18.000	3.000	13	18.000	5.000
39	18.000	21.000	16	18.000	2.000
14	18.000	4.000	10	18.000	8.000
11	18.000	7.000	14	18.000	4.000
11	18.000	7.000	14	18.000	4.000
9	18.000	9.000			
25	18.000	7.000	**Median**		**6.000**

deviation. A robust method is to consider median as the center. Absolute deviations from the median are then computed. Next we take the average value of these absolute deviations, that is, the median absolute deviation (MAD). The bug repair time data MAD value is 6.000. The calculation is shown in Data 3.3.

Sum of Squares and Variance

There is another way to avoid the sign problem. We can square the deviations and take the average. In some statistical contexts, we register an intermediate stage of computing the sum of squares. If two data sets have the same number of data points, the sum of squares can be used to compare dispersion. If the number of data points varies, we should take the average, known as *variance*.

For bug repair time, sum of squares and variation calculations are shown in Data 3.4.

Data 3.4   Sum of Squares and Variance of Bug Repair Time

Bug Repair Time Days	Mean Days	Squared Deviation Days	Bug Repair Time Days	Mean Days	Squared Deviation Days
16	20.517	20.400	25	20.517	20.100
23	20.517	6.167	20	20.517	0.267
45	20.517	599.434	17	20.517	12.367
20	20.517	0.267	13	20.517	56.500
13	20.517	56.500	13	20.517	56.500
13	20.517	56.500	13	20.517	56.500
58	20.517	1405.000	24	20.517	12.134
9	20.517	132.634	12	20.517	72.534
7	20.517	182.700	7	20.517	182.700
29	20.517	71.967	7	20.517	182.700
13	20.517	56.500	28	20.517	56.000
12	20.517	72.534	29	20.517	71.967
32	20.517	131.867	12	20.517	72.534
31	20.517	109.900	49	20.517	811.300
31	20.517	109.900	20	20.517	0.267
33	20.517	155.834	21	20.517	0.234
6	20.517	210.734	49	20.517	811.300
31	20.517	109.900	14	20.517	42.467
26	20.517	30.067	15	20.517	30.434
21	20.517	0.234	13	20.517	56.500
31	20.517	109.900	6	20.517	210.734
19	20.517	2.300	28	20.517	56.000
18	20.517	6.334	21	20.517	0.234
18	20.517	6.334	23	20.517	6.167
21	20.517	0.234	13	20.517	56.500
39	20.517	341.634	16	20.517	20.400
14	20.517	42.467	10	20.517	110.600
11	20.517	90.567	14	20.517	42.467
11	20.517	90.567	14	20.517	42.467
9	20.517	132.634	**Sum of Squares**		**7512.983**
25	20.517	20.100	**Variance**		**125.216**

The sum of squares is 7512.983. After deriving the average, the variance is found to be 125.216. Variance is a good measure for comparing data sets. However, the unit is days², a squared entity. One cannot make an intuitive assessment of dispersion as we are able to do with average absolute deviation.

BOX 3.2   ICEBERG ANALOGY

Data are like an iceberg. The peak contains only 15% of ice. The remaining 85% is beneath the water level, unseen by the onlooker. The unseen ice details could do great harm to ships. Likewise, the central values constitute just the tip. The real process behavior is in the spread of data. The true behavior of a process is understood when the spread is also recognized.

Standard Deviation

If we take the square root of the variance of bug fix time, we will obtain 11.190 days. This is the standard deviation, SD, of bug repair time, the most commonly used measure of dispersion. This is larger than the average absolute deviation.

The standard deviation is always larger than the average absolute deviation.

The exact formula for standard deviation, SD, has a small correction for sample size. Instead of using n as the number of data points, the exact calculation uses $n - 1$, the degrees of freedom; that is,

$$SD = \sqrt{\frac{\sum_{i=1}^{n} (x - \bar{x})^2}{n-1}} \tag{3.3}$$

The corrected value of standard deviation for bug repair time is 11.284.

Process dispersion can be defined in terms of standard deviation, sigma. It is a tradition dating back to the 1920s to take process variation as ±3 sigma. The normal distribution beyond ±3 sigma is disregarded. Mathematically speaking, the normal distribution runs from minus infinity to plus infinity. We trim the tails and take the span from −3 sigma to +3 sigma as the process dispersion. The trimming rules are associated with confidence level. The ±3 sigma trimming rule is associated with a confidence level of 97.3%.

BOX 3.3   THUMB RULES

With experience, people develop thumb rules about using dispersion measures. Although the rules depend on the individual person, here is an educative example. The following table shows three ways of applying dispersion.

SNo	Purpose	Range Considered	Confidence Level
1	Business decisions	Interquartile	50%
2	Process decisions	3–97 percentile	94%
3	Risk avoidance	Max–min	100%

Business decisions are customarily taken to accommodate IQR of variation. To accommodate more would need an unrealistic budget. Process decisions are made with expectations of reasonably stringent discipline. Risk avoidance involves understanding and accommodating extreme values. You can form your own rules of thumb to manage dispersion.

Skewness and Kurtosis

Pearson's Skewness

Skewness is a measure of asymmetry in data. Pearson's formula for skewness is based on the difference between mean and mode. If the difference is more, there is more asymmetry. The formula is given as follows:

$$\text{Skewness} = \frac{\text{Mean} - \text{Mode}}{\text{SD}} \tag{3.4}$$

Applying this formula to bug repair time data, we obtain a skewness of 0.666. The value is positive, indicating the presence of more large data. Data are said to be skewed to the right. If the skewness is negative, data would be negatively skewed, or skewed to the left.

If the mode is ill defined, then we can use the following modified formula based on the difference between mean and median:

$$\text{Skewness} = \frac{3(\text{Mean} - \text{Median})}{\text{SD}} \tag{3.5}$$

For bug repair time data, this formula yields a skewness of 0.669.

Bowley's Skewness

A robust estimate of skewness is based on quartiles and median. This is also known as *quartile skewness* or *Bowley's skewness*. The formula is given as follows:

$$\text{Bowley's skewness} = \frac{Q_3 + Q_1 - 2\text{Median}}{Q_3 - Q_1} \tag{3.6}$$

For bug repair time data, Bowley's skewness was calculated as 0.259. This value is much smaller than Pearson's value. Bowley's skewness is on a different scale; it varies from –1 to +1. Pearson's skewness varies from –3 to +3.

Third Standardized Moment

Skewness can be considered using the method of moments. Skewness is the third standardized moment. This convention is followed in Excel in the function SKEW that uses the following formula:

$$\text{Skewness} = \frac{n}{(n-1)(n-2)} \sum \left(\frac{x_i - \bar{x}}{s} \right)^3 \tag{3.7}$$

where n is the sample size, $\bar{x}$ is the sample mean, and s is the sample standard deviation.

It may be noted that in the formula, deviations are raised to the third power. Also, like in all skewness calculations, the value is normalized or standardized with a division by standard deviation.

The moment-based calculation skewness—using Excel function SKEW—for bug repair time data is 1.271. This is a more sensitive measure of skewness.

BOX 3.4   SKEWED LIFE

A good amount of software project data are skewed. Symmetrical and normal data are an exception. Data from simple processes show symmetry. Data from complex processes are skewed. Software development is certainly a collection of several processes and is expected to produce skewed data. If data collection is a process of observation, then we must recognize skew in data and learn to accept skew as a reality of life. The transformation of skewed data into symmetrical data is an artificial step performed often to apply some statistical tests. The untransformed raw data from software projects is often skewed. An outstanding example is the skew in complexity data. Another is skew in TAT data. In such cases, skew is the DNA of a process. Skew may restrict the application of several classic statistical methods while testing the data, but that is a secondary issue.

Kurtosis

The flatness of data is measured as kurtosis. The lower the value of kurtosis, the flatter the data distribution. There are different conventions in computing kurtosis. The Excel function KURT uses a formula for kurtosis given as follows:

$$\text{Kurtosis} = \frac{n}{(n-1)(n-2)(n-3)}\sum\left(\frac{x_i - \bar{x}}{s}\right)^4 - \left(\frac{3(n-1)^2}{(n-2)(n-3)}\right) \quad (3.8)$$

where n is the sample size, $\bar{x}$ is the sample mean, and s is the sample standard deviation.

The formula has been adjusted to make the kurtosis of normally distributed data equal to zero. The Pearson method of calculating kurtosis yields a value of 3 for normal distribution. If we subtract 3 from the Pearson result, we will obtain excess kurtosis. Hence, the Excel KURT formula gives "excess kurtosis," the value in excess of normal kurtosis. If this "excess kurtosis" value is positive, data are more peaked; if it is negative, data are broader.

Kurtosis for bug repair time data has been calculated. It is +1.676; hence, data are peaked.

Coefficient of Dispersion

The term coefficient is commonly used in algebra. The coefficient of a variable tells us the magnitude of the effect of the variable on the result. In metallurgy, the coefficient of expansion of metals can be used to calculate the expansion of metals. Here the coefficient is a metal property. The design of a coefficient of dispersion has a different purpose, although the connotations of the term are not entirely strange.

Coefficient of Range

The simplest coefficient of dispersion is the coefficient of range (COR), calculated as follows:

$$COR = \frac{(Max - Min)}{(Max + Min)} \tag{3.9}$$

For the bug repair data, COR can be computed as follows:

Max	58 days
Min	6 days
Max – Min	52 days
Max + Min	64 days
COR	0.8125 (dimensionless ratio)

Coefficient of Quartile Deviation

COR is based on extreme values and hence is not robust. Coefficient of Quartile Deviation (CQD) is based on quartiles and hence is not influenced by extreme values. The formula for CQD is given as follows:

$$CQD = \frac{Quartile_3 - Quartile_1}{Quartile_3 + Quartile_1} \tag{3.10}$$

For the bug repair time data, CQD is computed as follows:

Q_3	26.5 days
Q_1	13 days
$Q_3 - Q_1$	13.5 days
$Q_3 + Q_1$	39.5 days
CQD	0.342 (dimensionless ratio)

It may be seen that using quartiles gives a favorable value for the process of repairing bugs. CQD is much better than (smaller than) COR.

Coefficient of Mean Deviation

This is the ratio of average absolute deviation to mean value. For bug repair time data, the ratio is computed as follows:

Average absolute deviation	8.669
Mean	20.517
Ratio	0.423

Coefficient of MAD

This is the ratio of MAD to median. For bug repair time data, the ratio is computed as follows:

MAD	6
Median	18
Ratio	0.333

Coefficient of Standard Deviation

This is the ratio of standard deviation to mean. For bug repair time data, the ratio is calculated as follows:

SD	11.284
Mean	20.517
Ratio	0.550

This ratio is commonly known as *coefficient of variation* (COV). It can be expressed as a percentage. For bug repair time data, COV can be expressed as 55%. This is also called relative standard deviation (RSD).

Summary of Coefficients of Dispersion

For bug repair time data, the coefficient of dispersion has been studied using five different conventions, summarized as follows:

1. COR deviation	0.8125
2. CQD	0.342
3. Coefficient of mean deviation	0.423
4. Coefficient of median deviation	0.333
5. Coefficient of standard deviation	0.550

Higher values of this coefficient indicate problems because variation is seen as a risk. Estimates 1, 3, and 5 have been influenced by extreme values. Estimates 2 and

4 are robust, without any influence from extreme values. The true capability of the bug repair process is indicated by estimates 2 and 4.

Application Contexts

The statistic "dispersion measure" is most sensitive to context. Measures of dispersion can be applied in three prominent contexts: process control, experiments, and risk management.

Variation is unavoidable in software processes. In the manufacturing context, variation is the least in machine-controlled processes. Manual processes of hardware production have a few orders of magnitude more than variation. Software processes have several orders of magnitude more than variation. Software processes first have human variation; next most software processes are of a problem-solving nature and thus reflect variation in the complexity of the problem. Hence, Shewhart's common and special cause variations do not completely represent software process variation. In software processes, variation has subtler components, including genetic variation of agents and entropy of the problem scenario. We would rather attempt to understand variation before we classify variation in tune with the philosophy of Deming [1], which propounded that understanding variation is part of profound knowledge. Categorizing variation into types is divisive, whereas finding a numerical expression for dispersion is integrative. The numerical expression, robust enough to deal with nonnormal data, is MAD and can be used as a measure of process performance in performance scorecards. For instance, in the cases of bug repair, the following two values represent the process:

Median	18 days
MAD	6 days

If we study variation in experimental data, we will have a different context. In experiments, variation is treated as error. Truth is in the center. The standard deviation is a good measure to represent error. If the measured value is positive, we will benefit from using coefficient of standard deviation. When we do an experiment to measure productivity, we can express the experimental result as a mean ± % RSD (relative standard deviation or coefficient of standard deviation). For example, the mean productivity of 120 LOC per day ±30% RSD could be a good expression of experimental study.

Risk managers need a mathematical expression for variation. Of all the options, the standard deviation is a close enough approximation that works well for risk assessments.

The measures of dispersion given in this chapter provide a basic entry into the subject. For a cohesive understanding, variation should be modeled by methods given in Section II of this book.

In a Nutshell

Dispersion definitions used in chapter, in a nutshell, are presented as follows:

Measures of Dispersion

1. Range: maximum–minimum
2. Percentile range: 97th–3rd percentile
3. IQR: Q_3–Q_1
4. Average deviation: average deviation from mean
5. Average absolute deviation: average absolute deviation from mean
6. MAD: median value of absolute deviations from median
7. Sum of squares: sum of squares of deviations from mean
8. Variance: square of standard deviation
9. Standard deviation $SD = \sqrt{\dfrac{\sum\limits_{i=1}^{n}(x-\bar{x})^2}{n-1}}$

Nature of Dispersion

10. Pearson's skewness: $\text{Skewness} = \dfrac{3(\text{Mean} - \text{Median})}{SD}$

11. Quartile skewness: $\text{Bowley's skewness} = \dfrac{Q_3 + Q_1 - 2\text{Median}}{Q_3 - Q_1}$

12. Third standardized moment: $\text{Skewness} = \dfrac{n}{(n-1)(n-2)}\sum\left(\dfrac{x_i - \bar{x}}{s}\right)^3$

13. $\text{Kurtosis} = \dfrac{n}{(n-1)(n-2)(n-3)}\sum\left(\dfrac{x_i - \bar{x}}{s}\right)^4 - \left(\dfrac{3(n-1)^2}{(n-2)(n-3)}\right)$

Coefficients of Dispersion

14. Coefficient of range: $COR = \dfrac{(\text{Max} - \text{Min})}{(\text{Max} + \text{Min})}$

15. Coefficient of quartile deviation $CQD = \dfrac{\text{Quartile}_3 - \text{Quartile}_1}{\text{Quartile}_3 + \text{Quartile}_1}$

16. Coefficient of mean deviation: ratio of average absolute deviation to mean
17. Coefficient of MAD: ratio of MAD to median
18. Coefficient of standard deviation: ratio of standard deviation to mean

Case Study: Dispersion Analysis of Data Sample

This case study is from a support project. The data volume is pretty large. Around 15,000 incidents are logged every week. The turnaround time (TAT) of resolving the issues is taken for our study. We take a random sample of 30 data points from the database for dispersion analysis. Using a sample has its own risks: we may obtain a limited view of reality, and dispersion seen in the entire database may be quite large. However, analysis of a small sample is easy and provides a perspective and guidance for further analysis. The range of 30 data sample is 199 days, and it seems odd given the fact that there are tightly controlled service level agreements. The analyst remembers a 7-day service level agreement and is prompted to form clusters in the data. Three clusters emerge by visual analysis. The first cluster agrees with the memory recall of the analysts: the data are around 7 days. There seems to be a second cluster around 50 days. Two data points show extreme values of 80 and 200 days. A better evaluation of dispersion is possible if we use coefficients of dispersion and the analyst chooses the coefficient of MAD (CMAD). For the raw data, CMAD is high. After forming clusters and creating categories, the CMAD values become small and reasonable, as shown in Figure 3.1.

Category C seems to be special cases; perhaps those events were put on a low priority queue and were taken up very late. There is no information regarding this in the database; the only data logged in are time stamps of entry and release. A dispersion analysis of the data sample brought the problem in the database and prompted the analyst to create categories. Lessons learned from data sample dispersion analysis help in designing a framework for the big job: analysis of the total database.

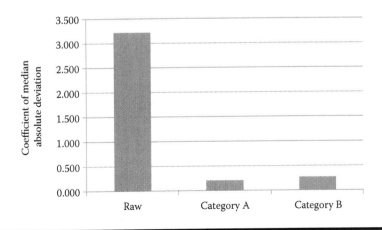

Figure 3.1 Dispersion analysis of data sample.

Review Questions

1. When will you use IQR instead of full range?
2. When will you use 3% trimmed range instead of full range?
3. When will you use range calculation based on standard deviation?
4. What is the benefit of using coefficients of dispersion?
5. Why do we prefer absolute deviations instead of plain deviations?

Exercises

1. Calculate skew in the data provided in Figure 3.1 in the case study using Pearson's and Bowley's method. Compare the results.
2. Use the bug repair time data from Data 3.4 and calculate the coefficient of standard deviation. What is your judgment on dispersion? Is it high or normal?

Reference

1. W. E. Deming, *Out of the Crisis*, MIT Press, Cambridge, Massachusetts, 2000.

Suggested Readings

Aczel, A. D. and J. Sounderpandian, *Complete Business Statistics*, McGraw-Hill, London, 2008.
Brown, S. *Measures of Shape: Skewness and Kurtosis*, Oak Road Systems, 2008–2011.
Heeringa, S. G., B. T. West and P. A. Berglund, *Applied Survey Data Analysis*, Chapman and Hall/CRC, April 5, 2010.
Lewis-Beck, M. S. *Data Analysis: An Introduction*, Sage Publications Inc., ISBN0-8039-5772-6, 1995.
Lunn, D., C. Jackson, A. Thomas, N. Best and D. Spiegelhalter, *The BUGS Book: A Practical Introduction to Bayesian Analysis*, Chapman and Hall/CRC, October 2, 2012.
Wuensch, K. L. *Skewness, Kurtosis, and the Normal Curve*, 2014, Available at http://core.ecu.edu/psyc/wuenschk/StatsLessons.htm.
Available at http://easycalculation.com/statistics/kurtosis.php.
Available at http://www.springerreference.com/docs/navigation.do?m=The+Concise+Encyclopedia+of+Statistics+%28Mathematics+and+Statistics%29-book62.
Available at http://www.uvic.ca/hsd/publicadmin/assets/docs/aboutUs/linksofinterest/excel/excelModule_3.pdf.

Chapter 4

Tukey's Box Plot: Exploratory Analysis

The Structure of the Box Plot

Box plot is easily the simplest and most widely used data analysis technique.

The origin of the box plot lies in the range plot. In the rudimentary version of the range plot, a line stretches between the minimum and the maximum values [1]. We can include markers in this range line to indicate central tendency. We can also annotate the line with markers to indicate standard deviation.

BOX 4.1 STATISTICAL THINKING

To achieve statistical thinking in engineering and management is our purpose. This involves thinking with data, perceiving central tendency and dispersion, and recognizing statistical outliers. These three aspects of statistical thinking are facilitated by the box plot. The central line in the box indicates central tendency, the median. Dispersion is shown in two levels of details: the length of the box is an indicator of dispersion in a broad business sense, and the whiskers indicate dispersion with more rigor and confidence level. Outliers, if any, are identified and plotted as points beyond the whiskers. To use the box plot is to practice statistical thinking. We can use the box plot effectively in management and engineering situations.

In its early form developed by Mary Spear in 1952, the box plot displayed the five-point summary of data [2]:

Median
Lower quartile
Upper quartile
Smallest data value
Largest data value

A box is made of median and quartiles; the box includes 50% of observations. The quartiles are the edges of the box called *hinges*. The whiskers are lines that begin at the hinges and end at the smallest and largest data values. The graph is known as the *box-and-whisker plot*, or simply the *box plot*.

The box plot has gone through several changes. A summary of the historical developments is presented by Kristin [3].

A simple but effective improvement of the box plot came from John Wilder Tukey, which made box plot a popular tool. Tukey modified the box plot and published it in *Exploratory Data Analysis* [4] in 1977. In the modern version, data fences are used. The whiskers do not stretch to the smallest or largest data values. The whiskers stretch out from the box only up to trimming points (or fences) that mark off outliers. The trimming rules have been empirically designed. The markers are 1.5 interquartile range (IQR) away from the box. Whiskers end at the points farthest from the box inside these markers. The markers provide a pragmatic way to find outliers. Aczel and Sundara Pandian, authors of an Excel tool to plot the box plot, refer to these markers as fences [5]. Besides these inner fences, the plot authors have introduced additional markers 3 IQR away from the box. These are referred to as outer fences. If data lie beyond the inner fences, they can be suspected as possible outliers. Data that fall outside the outer fences are definite outliers.

A typical box plot is shown in Figure 4.1. The following guidelines have been used in the construction of the graph:

Box central line	=	Median
Lower hinge (edge) of box	=	Quartile 1
Upper hinge (edge) of box	=	Quartile 3
IQR	=	Quartile 3 – Quartile 1
Right inner fence	=	Quartile 3 + 1.5 IQR
Left inner fence	=	Quartile 1 – 1.5 IQR
Right outer fence	=	Quartile 3 + 3 IQR
Left outer fence	=	Quartile 1 – 3 IQR

Software productivity data (lines of code/person day) are analyzed by this plot. The box is constructed from Quartile 1 (productivity = 8) to Quartile 3 (productivity = 34.5). Fifty percent of the data are inside the box. Hence, the core productivity is in

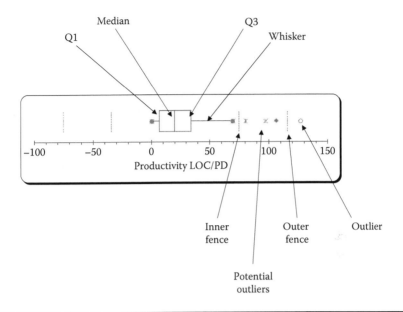

Figure 4.1   Box plot of software productivity.

this range. The left whisker reaches zero, whereas the right whisker reaches 70. The whisker ends represent a more complete range, beyond the core box. The whisker range has good news as well as bad news. The lower whisker is a serious concern; productivity values have dropped to zero. This could be a data error and might have to be cleaned out. We are doing exploratory analysis with the box plot, and we just make a note of this observation at the moment.

Then we find outliers. Those values between the fences are suspected outliers. Productivity values between 75 and 116 are perhaps not repeatable performance. Those values beyond the outer fence are definite outliers. They are, in a purely statistical sense, odd, untenable results. Perhaps those results might have had harmful side effects; the damage might have been done, and only a root cause analysis can tell.

Customer Satisfaction Data Analysis Using the Box Plot

In analyzing ordinal data, box plots are invaluable. Let us take the customer satisfaction (CSAT) index data from a development project. The data are shown in Data 4.1.

Data have been collected in a 0–10 scale. This scale fares better than the conventional 0–5 Likert Scale. The 0–10 scale has more granularity and less subjective error.

To understand the performance of the organization analysts, take the median if data were ordinal, although taking the mean is a common but mistaken practice.

**Data 4.1 Customer
Satisfaction Index Values**

5.0	5.8	7.4
6.5	7.6	8.7
6.6	6.0	8.3
9.1	7.2	7.7
5.9	5.5	7.1
5.8	6.8	6.5
7.7	6.3	8.1
7.5	5.9	7.0
5.0	6.3	6.4
5.9	8.5	4.6
7.1	8.2	8.0
7.6	6.0	7.3
7.9	6.2	5.9
7.2	4.7	6.9
7.8	5.3	6.7
4.7	6.9	8.1
7.0	6.7	7.1
3.1	7.6	7.0
5.1	8.2	6.8
4.5	7.2	6.6
8.1	8.6	6.9
5.9	6.9	6.6
7.8	7.7	

In this case, both mean and median provide nearly similar results. However, we prefer to use the median. The central tendency of CSAT index is shown as follows:

Mean   6.8
Median 6.9

This is compared with the organization goal, which happens to be 8.0. The obvious shortcoming is recognized, and future decisions are made to bridge the gap. This is the routine analysis.

Let us now try a box plot to display the CSAT data as shown in Figure 4.2.

We are able to make the following additional observations in the box plot.

1. The entire box is below the goal. This is a serious subject. The core process carrying 50% of performance results is below the mark.
2. There is an outlier with a value of CSAT index around 3.2. This is way down the track. If we apply the Kano model of CSAT, this score will run into deep dissatisfaction levels. Perhaps it is just short of customer fury.
3. Not a single event has reached the top score of 10. Customer delight seems to be an unattainable goal. To balance the outlier, we need at least a few delighters. To compensate one negative impression, we need to create ten positive impressions. The compensatory effort is missing.

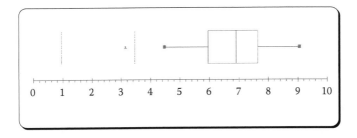

Figure 4.2   Box plot of customer satisfaction index.

Certainly, thinking with the box plot enables us to reason much more than working with the mean value of the CSAT index. Box plots make us see the problem in its entirety; this is a very valuable support.

Tailoring the Box Plot

The box plot is being widely applied in real life. We have seen the insights brought in by box plot to R&D scientists, project managers, business managers, quality managers, and data analysis in the software business. Even students in the middle grade in the United States are taught the box plot [6].

The box plot is being continuously refined. Tukey himself published variations in the box plot in 1978 [7]. Others have included frequency information in the box plot. Bivariate box plots called *bag plots* have also been tried out. People have proposed variants called *bean plots*. An analysis of the attempted improvements in the box plot may be found in the paper by Choonpradub and McNeil [8].

Attempts that have tried to pack more information into the box plot have failed. People prefer the simple uncluttered plain box plot.

Applications of Box Plot

Numerical quantities focus on expected values, graphical summaries on unexpected values.

John W. Tukey

Seeing Process Drift

If the median is not on a process target value, we can say that the process has drifted. The amount of drift can be easily seen if we draw a target line on the box plot. Figure 4.3 shows a box plot of bug repair time. The corporate goal is to fix bugs within a maximum of 16 days. The goal is marked on the box plot for easy interpretation.

Detecting Skew

The box plot is an eloquent way of expressing problems in process. One can see clearly if the process results are skewed. If the median is in the middle of the box, data are not skewed. If the median shifts to the right, data are left skewed. If the median shifts to the left, data are right skewed. Another sign of skew is the length of whisker. If the right whisker is longer, as seen in Figure 4.3, the process is skewed to the right.

Seeing Variation

The width of the box is a measure of process variation. Box width shows variation with 50% confidence level. The whisker-to-whisker width also expresses variation, perhaps with better clarity and more dramatic effect. The whisker-to-whisker range is an expression of variation with confidence levels more than 90%. In Figure 4.3, the whisker-to-whisker range is from 6 to 45 days. The variation is far in excess of what is anticipated.

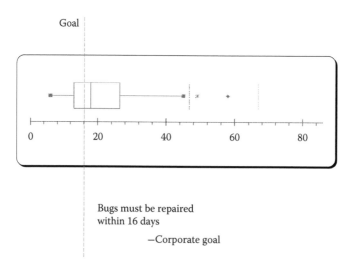

Figure 4.3 Box plot of bug repair time, days.

Risk Measurement

If we plot specification lines on the box plot, we can easily see the risk element in the process. If the entire whiskers stay outside the specification lines, the process is very risky. Risk is proportional to the portion of the box plot that stays outside process specifications. Bug repair, shown in Figure 4.3, surely has a schedule risk. We cannot quantify risk using a box plot, but we can qualitatively say the risk is very high. Risk management is one area where qualitative judgment is good enough and often more dependable than sophisticated quantitative analysis.

Outlier Detection

A very useful result from the box plot is the detection of outliers. The rules applied in the box plot do not assume any mathematical distribution function for the process. The box plot way of detecting outliers differs from the control chart way of detecting outliers. In control charts, we use the probability density function that corresponds to the inherent distribution of data. The box plot rules do not apply any distribution formula. The box plot uses a distribution-free judgment that is more universal and robust enough to engage all kinds of data.

Comparison of Processes

Box plots are used to compare process results. All the three elements of processes can be visually compared:

Central tendency
Dispersion
Outliers

This visual comparison performs the functions of three tests: *t* test for process mean, *F* test for process variation, and control chart tests for process outliers. This comparison is discussed later in this chapter.

Improvement Planning

Process improvement planning is well supported by box plot analysis. A box plot defines the problem with a picturesque essay of three dimensions of the process: central tendency, dispersion, and outliers. A box plot is an empirical problem statement. If we think that a well-defined problem is a problem half solved, then we stand to gain immensely by the box plot way of problem definition.

> An approximate answer to the right problem is worth a good deal more than an exact answer to an approximate problem.
>
> **John W. Tukey**

Box plots help us to identify and define the right problems.

The productivity box plot shown in Figure 4.1 highlights three opportunities for innovation:

1. *Removal of outliers*: This is the easiest innovation. There is no outlier in the lower side of the plot. That is good news. The outliers with higher values might appear as welcome outcomes. Here is the good old question of specifying an upper limit even for the better side of events. It may be suspected that extreme value of productivity is the result of a compromise, a slow acting fuse, that might show up later somewhere as an issue. Although we need to understand all the outliers, the outliers beyond the outer fence may be studied in detail. The presence of more outliers on the right side also indicates the possible existence of a tail or skew in the distribution.
2. *Shifting the median toward higher levels*: This means the expected value of productivity can be improved.
3. *Reduction of IQR as well as whisker-to-whisker width*: Process variation, depicted both by the IQR and whisker-to-whisker width, can be reduced to minimize variation.

The three innovations could coexist. Improving the median may be accompanied by reduction in outliers, and vice versa. It is a good strategy to take up one at a time, in the previously mentioned order, and take the beneficial side effects in the other two.

Core Benefits of Box Plot

Tukey's box plot contains sufficient statistical strategies and yet retains its intended simplicity. Many attempts are being made to enhance the information content in box plots and make them colorful as well. We focus on the simple box plot in this chapter and find that it has great potential. The box plot can be applied to the following:

■ Provide a visual summary of data
■ See process variation
■ Detect outliers
■ Detect skew in data

- See process drift
- Compare processes
- Plan process improvement

Twin Box Plot

Let us take the case of reestimating software development effort. Teams are reluctant to do a second estimate and are in a hurry to move forward with development. However, it is a best practice to do a second estimate after a fortnight into the project when many project details become visible. We get to know the requirements better, teams communicate better, risks are seen with clarity, and we are enlightened by the early lessons. The second estimate is expected to be more accurate. We wish to compare the second estimates with first estimates and study the improvement.

The box plot can be eminently used to compare the two results. In Figure 4.4, a twin box plot is shown comparing two sets of effort variance data.

The twin box plot offers what might be called a *visual test*, a preliminary analysis before we start rigorous tests. Visual judgment of the following can provide vital clues regarding the differences between two results:

1. Is there a difference between the whisker-to-whisker widths?
2. Is there a difference between the box widths?
3. Is there a relative shift in the position of the median?

If the answer is yes to any one of these questions, we need to take a deeper look at the box plots. Sometimes the presence or absence of outliers could make a difference. Sometimes the skew of the median line inside a box could provide a clue.

If the difference is significant, the boxes in the two plots may be completely disjointed. They may not overlap. Using the box plot representation, it is rather easy to see if the new result is different from the old.

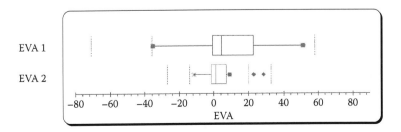

Figure 4.4   Comparison of two estimates using box plots.

If results due to innovation show improvement, one or more of the following visual clues may be present:

- The overall length of the box plot would have decreased
- Outliers might have disappeared
- The central line might show a favorable shift
- The box might have shrunk
- The box might be relocated in a favorable region
- The unfavorable whisker might have diminished

If an improvement is not visible in a box plot, it may not be an improvement in the first place. The question of looking for significance does not arise.

> *However, in most cases, people take pains to go through lengthy procedures to execute significant tests to check differences, without box plot visual checks. In some cases even after box plot rejections, people go through the ritual of significance tests.*

Holistic Test

The twin box plot test is a holistic approach; it can compare two populations (two groups) in a complete balanced fashion that no other test can offer. The price we pay for completeness is loss of rigor. It so happens that rigorous tests have narrower scope than robust tests; approximate analysis can sweep more terrain than precise analysis. We need such a holistic test before we go into more sophisticated tests.

The twin box plot shown in Figure 4.4 offers a holistic comparison described in the following paragraphs.

First, it compares the median values. The median of the first estimate is 4.67%, and the median of the second estimate is 1.27%. Comparing medians is more robust than comparing means, which makes sense even with nonnormal data. This is a comment on central tendency.

Then dispersion is compared at two levels; the first IQR is 23.45 and the second and improved value is 8.54. It is evident that the core of the estimation process covering 50% of results shows less dispersion—an order of magnitude less. The new dispersion is one-third the old. The old whisker-to-whisker range is 86.48, whereas the new whisker-to-whisker range is 20.32, four times less. It is evident that the dispersion is reduced in the new estimation technique; it is more reliable. The box plot provides an order of magnitude test before we resort to *p* values for judgment.

The box plot identifies outliers in the second group; not every estimate has been well performed. The best practice must spread. The second process has philosophical problems called *statistical outliers*. However, in a practical sense, even the outliers are better than the first process.

Application Summary: Twin Box Plot

- The twin box plot is a qualitative test and should be performed before any hypothesis test.
- The only way to compare overall performance of data sets is the twin box plot.
- We can compare the following aspects of process using twin box plots:
  - Quartile-to-quartile distance (IQR)
  - Whisker-to-whisker distance
  - Range
  - Median (central tendency)
  - Outliers
  - Skew
- Each comparison can provide a unique clue about difference in processes.
- We can use a rule of thumb: if boxes overlap, there is no significant shift in central value.
- After seeing the twin box plot, we can decide which confirmatory test must be performed.
  - If there is shift in central value, confirm it with a t test.
  - If dispersion is different, confirm it with an F test.
  - If outliers are present, cross check them with a control chart.
- We can take preliminary decisions with the box plot, followed by confirmatory tests to make the final decision.
- When data are nonnormal, twin box plots provide more reliable clues than conventional tests.

BOX 4.2   EVALUATING IMPROVEMENT

The need to evaluate that improvement occurs more often than we think in software projects. In the very first place, we collect data because we wish to improve performance. We are thus made to check if performance has really improved after data collection and reporting. To do this, we need two sets of performance data, before and after improvement. Then we just have to prepare a twin box plot and compare the results, as described in this chapter. There are other circumstances when we consciously improve performance through six sigma and lean; once again, we can use box plots to compare results before and after improvement. Sometimes we may do special experiments and invariably use the box plot to portray data using box plots as evidence for improvement. Box plots are widely used as graphical companion to experiments. High maturity in software engineering involves continual improvement, and the box plot is a very valuable tool.

Case Study 1: Business Perspectives from CSAT Box Plots

This case study is about managing CSAT across a large organization with four strategic business units (SBUs). The annual average CSAT index of the organization has been computed as 3.217, which is far below the target of 4 in a Likert scale of 1–5. The CSAT data are obtained by a survey of the overall satisfaction of customers. The calculation of the average of ordinal data is a subject of ongoing debate. Strictly speaking, average is meaningless in ordinal data, but average is taken as an effective indicator. It is easier to estimate and report. If we use box plots instead of mean, we tend to see more details of CSAT. If we plot separate box plots for different SBUs, we get more information and an easy intercomparison, as shown in Figure 4.5.

CSAT Analysis across SBUs

In one glance, we are able to take in several details of CSAT. The linear structure of box plot accommodates several box plots in one chart. In Figure 4.5, there are lower whiskers; the lower whiskers touch the floor level, particularly in SBU 1 and SBU 3. These lower whiskers are the real problems; customers tend to remember negative results longer. If Kano's model of CSAT can be used, the lower whiskers fall in the zone of asymptotically crashing dissatisfaction.

The key message of CSAT Box Plots is not in the central values but in the lower whiskers.

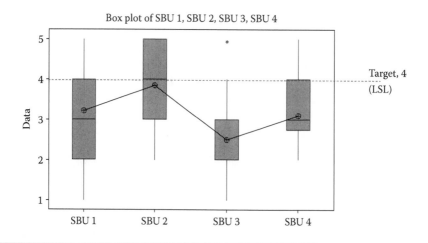

Figure 4.5   Customer satisfaction analysis across SBUs using box plots.

The chart shows SBU 2 to be outstanding. The box reaches the maximum value, providing customer delight. In SBU 3, customer delight is seen as a rare achievement and not a repeatable result. The chart is a sort of control chart on CSAT across the organization. Target 4 can be interpreted as the lower specification limit; and the chart provides a clear perspective of how CSAT performance meets target.

Case Study 2: Process Perspectives from CSAT Box Plots

In another case study, we show how multiple dimensions of CSAT can be tracked using box plots. The CSAT survey has captured customer responses to several other dimensions of CSAT:

Engineering (ENGG)
Communication (COMM)
Time (TIME)
Price (PRICE)
Responsiveness (RESP)
Quality (QUAL)

These selected six dimensions, or CSAT attributes, captured by the survey indicate customer responses and provides opportunities for improvement to the software development organization. The six box plots are available in a single chart as in Figure 4.6.

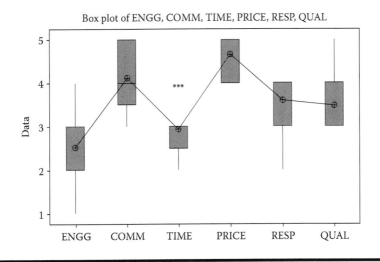

Figure 4.6 Customer satisfaction analysis across attributes using box plots.

BOX 4.3 THE BOX OF THE BOX PLOT

The box plot has a lean structure. It is remarkably simple and uncluttered. The earliest version of the box plot was a straight line. Tukey added the box. The box achieves its purpose by dominating the plot. This is an intended domination. The box has the median and contains 50% of central evidence. The first glance makes us recognize the box and other details are subdued; the box emerges as the primary message. This helps managers to grasp the essential behavior of processes sans the secondary details. Dispersion beyond the box is considered secondary. Outliers are tertiary. Moreover, the box is plain and unpopulated. It is just an outline drawing. For quick decisions regarding budgeting, the box is all we need to consider. For systematic process management, we consider the whiskers. For strategic risk management and problem solving, we consider the outliers. The structure of the box plot helps with this progression of management decisions.

CSAT Analysis across Attributes

The chart provides a very easy comparison that we can quickly navigate through.

Performances in ENGG and TIME have earned the lowest scores. The lower whisker in ENGG touches the floor and provides a red alert to the organization. COMM and PRICE have earned the best scores, assuring customer delight from within the "box" area. If we take the target as 4, then RESP and QUAL still need improvement. The box plots provided very useful information graphically.

Review Questions

1. What are the elements in a box plot?
2. What are fences?
3. How is the length of a whisker calculated?
4. How are the hinges calculated?
5. How robust are the rules used in detecting outliers in the box plot?

Exercises

1. Draw a box plot using the data provided in Data 4.1, using the macros provided in Refs. [4] and [5], and interpret the same. You can also download free Excel box plot plotters from the web.
2. Draw a box plot of lines of code developed by yourself or various objects. See how the box plot helps in statistical thinking.

3. Draw a box plot of defects based on the following module data. Interpret your findings.

26
23
21
18
18
18
14
14
13
13
12
12

4. Compare the following two sets of rework efforts during testing using two box plots. Interpret your graphs.

Set 1	Set 2
12	10
12	10
0	9
5	9
7	9
8	7
9	7
0	7
	7
	6
	6
	6
	5
	4
	4
	4
	4

References

1. K. W. Haemer, Range-bar charts, *American Statistician*, 2(2), 23, 1948.
2. M. E. Spear, *Charting Statistics*, McGraw-Hill Book Company, Inc., New York, 1952.
3. K. Potter, *Methods for presenting statistical information: the box plot*, in *In Visualization of Large and Un-structured Data Sets*, GI-Edition Lecture Notes in Informatics (LNI), H. Hagens, A. Kerren and P. Dannenmann (eds.), vol. S-4, pp. 97–106, 2006.

4. J. W. Tukey, *Exploratory Data Analysis*, Addison Wesley, Sydney, 1977.
5. A. Amir and J. Sundara Pandian, *Complete Business Statistics*, McGraw-Hill, London, 2008.
6. A. Bakker, R. Biehler and C. Konold, *Should Young Students Learn about Box Plots?* Curricular Development in Statistics Education, Sweden, 2004.
7. R. McGill, J. W. Tukey and W. A. Larsen, Variation of box plots, *American Statistician*, 32(1), 12–16, 1978.
8. C. Choonpradub and D. McNeil, Can the box plot be improved? *Songklanakarin Journal of Science and Technology*, 27(3), 649–657, 2005.

METRICS

Deriving metrics from data is the key to observation and understanding. Metrics endow people with objective observations and deeper understanding. In Section II of this book, Chapter 5 is devoted to deriving metrics from data to gain this advantage.

Common metrics for all life cycle phases are not an effective approach. Compartmentalizing metrics helps. A simple way of doing this is to separately discuss development metrics, maintenance metrics, and test metrics. Chapters 6, 7, and 8 are devoted to these three categories of metrics.

The advent of agile methods has redefined the way life cycles are managed and measured. Chapter 9 addresses agile metrics.

Chapter 5

Deriving Metrics

Creating Meaning in Data

Direct observations are also called *base measures* or *raw data*. Such data are either entered in the computer by people or recorded automatically by machines. Automated data collection is gaining currency. When people enter data, they have a chance to see what they are keying in and validate the data. Data caught by machines are not immediately seen by people; such automatic data are visited during report generation or process analysis.

Derived measures are computed from base measures. Derived measures are known by several names. Two of these names are significant: key performance indicators and metrics; we use the term metrics. Errors in base measure propagate into metrics. In a broader sense, metrics also constitute data. However, metrics carry richer information than base measures. We create meaning in data by creating metrics.

Deriving Metrics as a Key Performance Indicator

Measurement is essentially a mapping process. The primitive man counted sheep by mapping his cattle, one to one, to a bundle of sticks. If a sheep is missing, it will show up as a mismatch. Word was not yet invented, but there was mapping all the same. A similar mapping is performed for function point counting; measurement is seen as a "counting process," a new name for mapping. The mapping phase in measurement is well described in the COSMIC function point manual. With the help of language, we have given a name for what is counted—software size. With the help of number theory, we assign a numerical value applying rules.

In a similar manner, we count defects in software. Here the mapping is obvious. The discovery of defect is conducted by a testing process. Each defect is given a name or an identification number. The total number of defects in a given module is counted from the defect log.

Size is a base measure. Defect count is another base measure. The ratio of defects to size is called *defect density*. It is a derived measure, a composite derived from two independent base measures. It denotes product quality.

Productivity is another example for a derived measure.

Measures are directly mapped values. Metrics are meaningful indicators constructed against requirements.

> *A "measure" refers to directly mapped value.*
> *A "metric" refers to a meaningful indicator.*

Technically speaking, size is a measure, and complexity is a metric. Arrival time is a measure, and delay in the arrival is a metric.

Metrics carry more meaning than measures. Hence, metrics are "meaningful indicators."

Estimation and Metrics

A few metrics such as effort variance and schedule variance are based on estimated and observed values. For instance, the metric effort variance is defined as follows:

$$\text{Effort variance \%} = 100 \times \frac{(\text{Actual effort} - \text{Estimated effort})}{\text{Estimated effort}}$$

This metric truly and directly reflects any uncertainty in estimation.

> *Accurate measurement combines with ambiguous estimation to produce ambiguous metrics.*

Measurement capability and estimation system support each other. They are similar in so much as both are observations. Metrics measure the past and the present; estimation measures the future.

Paradigms for Metrics

What's measured improves.

Peter F. Drucker

"Measure what matters" is a rule of thumb. We do not measure trivial sides. We do measure critical factors. Intrinsic to this logic is an assumption that having metrics is an advantage; we can improve what we do. The balanced score card measures performance to achieve improvement. Areas for improvement are identified by strategic mapping. Loyal followers of the balanced score card way use this method to improve performance through measurements.

Another paradigm for measurement can be seen in quality function deployment (QFD). This is an attempt to measure what's and how's. The QFD structure and the associated metrics have benefitted several organizations.

Capability maturity model integrated (CMMI) suggests measurement of every process at each level of process maturity. The list of metrics thus derived could be comprehensive. The goal question metric (GQM) paradigm is suggested to select metrics at each level.

ITIL suggests measurements to improve service quality.

ISO 9000 indicates the measure–analyze–improve approach. It protects quality of data by meticulous calibration of measuring devices.

The Six Sigma initiative suggests metrics to solve problems. It has a measure phase, where $Y = F(X)$ is used to define X (causal) metrics and Y (result) metrics.

In the lean enterprise, wastes and values are identified and measured to eliminate waste.

In clean room methodology, software usage is measured and statistically tested. Reliability metrics are used in this case.

In personal software process (PSP), Humphrey proposed a set of personal level metrics. The choice was based on the special quality and attributes of PSP.

Barry Boehm uses a narrowed down set of metrics to estimate cost in his cost construction model (COCOMO). COCOMO metrics have created history by contributing to estimation model building.

A metric design follows the framework used for improvement. There are many frameworks and models for achieving excellence. Metrics are used by each of them as a driver of improvement. The system of metrics easily embraces the parent framework.

GQM Paradigm

Most metrics are naturally driven by strong and self-evident contexts. In special applications such as breakthrough innovation, model building, and reliability research, we need special metrics. We are very anxious that metrics carry a purpose. Special initiatives and hence special metrics should still connect with business goals. The tree that makes the connection is the GQM paradigm [1].

The GQM paradigm is an approach to manage research metrics and hence is more effective in problem solving and model building. It is not so influential in driving the five categories of industry metrics mentioned earlier.

Software Engineering Institute (SEI) introduced the GQ(I)M paradigm [2] as a value adding refinement to the GQM paradigm. GQ(I)M uses Peter Singe's mental models to drive metric choice and indicators to convey meaning. GQ(I)M certainly has helped to widen the reach of GQM.

BOX 5.1   FLYING A PLANE AND GQM

Even the simplest propeller airplane would have meters to measure altitude and fuel. These measurements are intrinsic to the airplane design. The meters are fitted by the manufacturer and come with the airplane as basic parts of the airplane. One cannot fly without altitude, speed, and fuel level metrics. Flying a plane without altimeter is unthinkable. A plane without a speedometer is unrealistic. A pilot cannon make decisions without a fuel indicator. These metrics are not "goal driven" and certainly not business strategy driven but are driven by design requirements. One can think of purposes for each metric, but these purposes are not derived from business strategies and business goals; these purposes are implicitly inherent in product engineering. Whether there are goals or not, these meters will be fitted to the plane, almost spontaneously, like a reflex action triggered by survival needs. There are no options here. These metrics are indispensable and obvious. One does not need a GQM approach to figure them out.

Hence, cost, schedule, and quality metrics are also indispensable in a software development project. These are not "goal driven" but are based on operational needs. One does not have a choice.

Difficulties with Applying GQM to Designing a Metrics System

First, the intermediate stage (question) in the GQM paradigm is not very helpful. We simply map metrics to goal. The mapping phase in COSMIC size measurement is a great illustration.

Second, while applying GQM, people tend to start with corporate goals and attempt to drill down to metrics. This often turns out to be a futile attempt. Large organizations have spent days with GQM, getting lost in defining goals, subgoals, and sub-subgoals. All these goals go into goal translation and could never make it to metric definitions in a single workshop. Rather, we would first derive performance goals from business goals using a goal tree. This is a goal deployment, a leadership game. Designers of metrics should discriminate metrics mapping from goal deployment. Designers of metrics should pick up selected performance goals and map them to metrics.

Third, some metrics are specified by clients. Customer-specified metrics seem to run very well in organizations. Data collection is smooth. There is no need for a separate mechanism to identify and define these metrics.

Fourth, some metrics are driven by requirements. If requirements include that the software must be developed for maintainability, there is a natural metrics associated with this requirement, that is, the maintainability index. Meeting both functional and nonfunctional requirements might need metrics support. Thus, the recognized performance targets easily and organically map into performance metrics. One need not apply GQM and make heavy weather.

Fifth, metrics are often constructed to serve information needs in the organization. Hence, management information systems (MIS) automatically capture many metrics. These metrics are an inherent part of MIS. Metric teams have to extract metric data from the MIS or ERP systems. These metrics do not follow the GQM road.

Sixth, some metrics are derived from operational needs, for example, schedule variance. Such needs are compelling, and one does not have a chance to exercise options. When the needs are clearly and decisively known, we need not rediscover them by GQM.

Seventh, even in the Six Sigma way of problem solving, where the Y and X variables are defined to characterize the cause-and-effect relationships that constitute the problem, metrics are derived by mapping through a cause–effect diagram. The selection of the problem to be solved is a goal-driven process, but deriving the variables (metrics) is conducted through causal mapping, not GQM.

Need-Driven Metrics

It is our finding that successful metrics are driven by transparent needs. The link between metrics and needs must be organic, spontaneous, and natural. The bottom line:

> *If we can do without metrics, we will do without metrics. We use metrics only when they are needed.*

The system of assigned goals, personal goals, and all the subgoals finally boil down to performance goals that reflect the pressing needs of the system. Once a metric connect with needs, it works.

The connection between needs and metrics must be concrete, spontaneous, transparent, and direct. Hence, mapping is the preferred connecting mechanism. Using "questions" is too verbose to be of practical value.

Mapping is a better connector than question.

A more serious concern would be to obtain commitment from agents to support metrics. A need-based system enables commitment harder than inquiries and questions.

There is a difference between goal and need.

Goal: the end toward which effort is directed
Need: a condition requiring supply or relief

Goals are complex; they consist of interconnected layers and are influenced by "personal goals" and "self-efficacy." Goals have multiple dimensions, are likely to drive discussions of metric design into inconclusive divergence, and are correspondingly undesirable to rely on for deriving metrics. Needs have a simple structure and are well defined with greater degree of objectivity in software development projects.

BOX 5.2 OLYMPIC RUNNER

Time in a school's final sprint competition is measured using an analog stopwatch. In interschool competitions at the state level, time is measured more precisely using a digital stopwatch. In Olympic sprints, time is measured by laser systems controlled by computers to the precision of a millisecond. As the capability of running improves, the precision of measurement also is improved. Likewise, when software engineering practices become more mature, metric capability also is improved. The quality of metric data also is improved. Metrics and maturity go together.

Meaning of Metrics: Interpreting Metric Data

We define a metric by defining the relationship the metric has with raw data. Metric definition inheres the meaning of that metric. The defining equation is more significant than the name we give to a metric. Names could mislead, but definitions do not. It is good to recall metric definitions, even if obvious, before beginning interpretation.

Having recalled the metric definition, we now look at metric data. It is better to work with a data table that shows the raw data and the derived metric in separate columns. Visibility into basic observations helps in getting a detailed understanding of metrics. In one column, we can have the time stamp of data. If data fall into categories, it is better to include the category name in one column. Occasionally, we may have to allocate additional columns to accept further categorization schemes.

Next we should construct a box plot of metric data and analyze the statistical nature of data. The box, whiskers, and outliers seen in the box plot must be understood and explained. Questions regarding the stability and the reasonable dispersion of the metric must be addressed. We can support this enquiry with a descriptive statistics analysis.

Data have intrinsic meaning that can be seen by applying statistical, engineering, and management perspectives.

The engineering and management understanding of the "variable" the metrics denote should throw more light into the box plot. The box plots must be compared with expected behavior.

Finally, we should mark the performance goal line on the box plot and relate data behavior to the goal. Figure 5.1 shows the box plot of productivity metric with the performance goal marked. This performance goal represents the project objective and not the business goal because this metric is treated as a project metric in our example.

Meaning of metric is seen relative to the operating goal.

How data relate to performance goals is what we now learn from data. Does the box include the performance goal line, or has the performance goal slipped away from the box area? Or, in the worst case, has the performance goal (quantitative target) drifted far and gone over to the whiskers?

We can now develop the various interpretations and come to some conclusions. For instance, interpreting the productivity metric of Figure 5.1 has led to the following interpretations and conclusions:

Most of the results, denoted by the box and whiskers, fall short of the performance goal or target value.

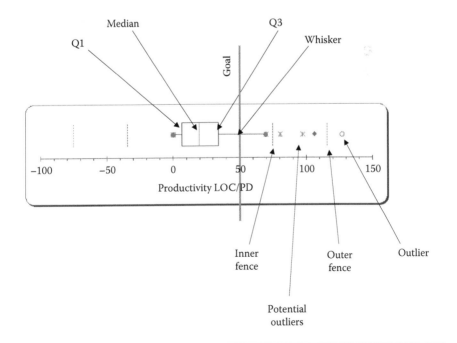

Figure 5.1   Box plot of software productivity.

Our Categories of Metrics

There are different ways to classify metrics. We follow the five classifications of data mentioned in Chapter 1 and work with the corresponding five types of metrics:

1. Business metrics
2. Project metrics
3. Process metrics
4. Subprocess metrics
5. Product metrics

Some of these categories could overlap depending on the definitions used in the organization. The same metrics could be reused in a different context under a different category. For example, reuse is generally regarded as a process metric. However, it can be used as a product metric to evaluate product behavior, as a project metric to understand time saved by reuse, and as a business metric to estimate profit generated by reuse. Context influences categorization.

Business Metrics

These are defined and deployed to implement business strategy in the organization. The metrics are strategic in nature. A popular example of the use of business metrics may be seen in the balanced score card framework of Kaplan et al. [3]. In this framework, metrics are identified under four categories: finance, process, learning and growth, and customer. Kaplan promises progress through measurements. These metrics interact and from a system with cause–effect relationships. Deriving business metrics from business strategy is very similar to the approach in the GQM paradigm of Victor Basili [1]. However, instead of using a translation tree, metrics are mapped to strategy.

 Business metrics are driven by strategy and situation.

Project Metrics

Managing a software project needs some minimum metrics such as effort, schedule, quality, productivity, and customer satisfaction (measured through surveys).

Project metrics are based on project requirements and customer requests. These are driven by project scope. The list of project metrics is extended to accommodate customer requirements such as reliability and maintainability. Some customers show keen interest in knowing product quality, down to the module level. They seek reports on module quality and module performance during testing. Project metrics also have to satisfy information to be communicated to stakeholders. For example, resource utilization can become an important metric sought by human resources and finance departments. Server downtime is a metric that would help the facilities management department. Projects also measure risk by internal surveys.

In general, project metrics are requirement driven. They are to be specified during scope definition. Project metrics also serve as information devices and assist in project communication.

> ### BOX 5.3   ANALOGY: CONSTRUCTING A BUILDING
>
> Software development is analogous to constructing a building in some ways. The natural metrics in a civil construction are the size of the building, the grade of quality of interior and materials used, and the reliability of the building against natural forces such as wind and rain. Similarly, in constructing a software product, the basic metrics are size, complexity, and reliability. In a construction business, routinely maintained design and construction logbooks show these metric data. In software projects, these metrics are slowly being accepted; a few organizations maintain these records, and many others still wait.

Process Metrics

Every process presents opportunities for metrics at the input, process, and output stages. Such a process metric is used first to understand, then to control the process, and later to improve the same.

Use metrics to understand, control, and improve.

Software delivery is made through a chain of processes, and one or more metrics can be used to control and improve each process stage in the chain. The total number of process metrics is equal to or more than the number of stages. Metric choice is driven by process control needs. This choice of metrics is further influenced by the life cycle model used. For example, in the waterfall life cycle, metrics such as requirement stability, design complexity, code size, and test effectiveness can be used to address control needs in each development phase.

Subprocess Metrics

A more detailed process control takes us to the subprocess. For example, a process, such as design, can be broken down to manageable subprocesses such as design, design review, and redesign. The metrics for controlling these subprocesses are design effort, design review effort, and design rework effort. Subprocess metrics present a management challenge because these metrics get closer to the individuals who fear exposure and hence refrain from sharing data. As Deming said, we need to "drive out fear" if we want to collect subprocess metric data.

Product Metrics

Software product structure and performance are monitored through product metrics. For example, we can measure software code size by lines of code, structure by function point, quality by defect density, and reliability by mean time between failure. In the design stage, we can measure structure by function point and size by number of design pages. In the requirement stage, we can measure size by number of features. Product metrics can be used to monitor "product health."

> ### BOX 5.4 EARTHQUAKE
>
> Earthquake prediction uses esoteric metrics. For example, the rise and fall of water levels in deep wells in China are measured; elsewhere, in a scientific approach, weak magnetic signals are monitored using sophisticated equipment to measure seismic activity. Seismic vibrations are picked up along fault lines and analyzed to predict earthquakes. All these metrics are special, expensive, and based on geological models. Likewise, software reliability prediction needs special metrics. Metric choice depends on the reliability model selected for prediction. These metrics are not routine but special, and they belong to the category of "model-driven metrics." They have to be specially designed and deployed, and the extra cost and effort should be approved by business leaders.

 ## Case Study: Power of Definitions

The definition of metrics precedes metric data. Metric definitions shape our approach to engineering before data validates our approach. The names of metrics with their definitions are part of the engineering and management vocabulary. Richer vocabulary reflects richer practices. Asking how many metrics we need is like asking how many words we need to converse effectively.

A few words are enough to exchange pleasantries,
A hundred are enough to manage simple conversations,
A thousand makes one an expert communicator, and
Many more are used by professionals.

On the other side of it, how effectively we use the few metrics is more important than the vocabulary volume. When managers use metric vocabulary in day-to-day dialogues with team members, a new culture can be created. Here is the case of one manager who wanted to make his software more maintainable. He chose to ask his programmer for a definition of maintainability; the response was swift, the programmer found out, from the literature, a formula for maintainability index. The manager left the subject at that and did not press for either the use of this index or the maintainability data on the code. The very definition of the maintainability index triggered a chain of responses from the programmer, from a realization that maintainability is important to improve code maintainability. Numbers were gathered only later, in subsequent trials. The numbers were not shared with others. All that the programmer needed was direction, and the manager showed that he was a great leader by giving the direction.

Lessons:

Metrics thrive under great leadership.
Some metrics are very personal.

BOX 5.5   MEET THE EXPERT—WATT S. HUMPHREY

Watt S. Humphrey (1927–2010), known as the father of software quality, was born in Battle Creek, Michigan. He enlisted in the Navy at 17 years of age to help fight in World War II. After his enlistment was up, he enrolled in the University of Chicago where he graduated with a bachelor's degree in physics. He earned his master's degree in physics from the Illinois Institute of Technology and then a master's degree in business administration from the University of Chicago.

He started his career with Sylvania in Boston and then moved to IBM, where he rose through the ranks to become director of development and vice president of technical development. In that job, he supervised software development in 15 laboratories that were spread out in seven countries. There were 4000 software engineers working under him.

At 60 years of age, when many people are thinking of retiring, Mr. Humphrey embarked on a new career at Carnegie Mellon University, where he established the software process program that instilled a discipline to software development.

His colleague, Anita Carleton, the director of the Carnegie Mellon Software Engineering Institute's Software Engineering Process Management Program, said that before Mr. Humphrey came along, software engineers created programs by coding and testing. He changed the culture of the discipline to develop a more systematic approach to planning, developing, and releasing new software.

His work earned him the National Medal of Technology, which was presented to him by President George W. Bush in 2005.

Review Questions

1. What are the five categories of metrics used in software projects?
2. What is the GQM paradigm? What are its limitations?
3. What is the GQ(I)M paradigm? What are the advantages of GQ(I)M over GQM?
4. What is the primary motivator of project metrics?
5. How are process metrics selected?

Exercises

1. Develop a metric plan to manage software testing.
2. Develop a metric plan to control design complexity.
3. Develop a metric plan to control code quality.
4. Develop a metric plan to control requirements volatility.
5. Develop a metric plan to manage software maintenance.

References

1. V. R. Balili and G. Caldiera, *Goal Question Metric Paradigm*, John Wiley, 1994.
2. R. E. Park, W. B. Goethert, W. A. Florac, *Goal Driven Software Measurement— A Guidebook*, SEI Handbook CMU/SEI-96-HB-002, 1996.
3. R. S. Kaplan and D. P. Norton, *The Balanced Scorecard: Translating Strategy into Action*, Harvard Business Review Press, Harvard College, 1996.

Suggested Readings

Aczel, A. D. and J. Sounderpandian, *Complete Business Statistics*, McGraw-Hill, London, 2008.

Austin, R. D., *Measuring and Managing Performance in Organizations*, John Wiley & Sons, Inc., Hoboken, NJ, June 1, 1996.

Defeo, J. and J. M. Juran, *Juran's Quality Handbook: The Complete Guide to Performance Excellence*, 6th ed., McGraw-Hill Professional, London, 2010.

Juran, J. M. and A. Blanton Godfrey, *Juran Quality Hand Book*, 5th ed., McGraw-Hill Professional, London, 2010. Available at http://www.pqm-online.com/assets/files/lib/juran.pdf.

Kan, S. H., *Metrics and Models in Software Quality Engineering*, Addison-Wesley Longman Publishing Co. Inc., Boston, 2002.

Kitchenham, B., *Software Metrics: Measurement for Software Process Improvement*, Blackwell Publishers, Inc., Cambridge, MA, 1996.

Laird, L. M. and M. C. Brennan, *Software Measurement and Estimation: A Practical Approach*, Quantitative Software Engineering Series, Wiley-Blackwell, 2006.

Parasoft Corporation, *When, Why, and How: Code Analysis*, August 7, 2008. Available at http://www.codeproject.com/Articles/28440/When-Why-and-How-Code-Analysis.

Parmenter, D., *Key Performance Indicator*, John Wiley & Sons, 2007.

Pearson, *Software Quality Metrics Overview*, Metrics and Models in Software Quality Engineering, 2nd ed., 2002. Available at http://www.pearsonhighered.com/samplechapter/0201729156.pdf.

Software Metrics, Available at http://www.sqa.net/softwarequalitymetrics.html.

Software Process and Project Metrics, Chapter 4, ITU Department of Computer Engineering–Software Engineering. Available at http://web.itu.edu.tr/gokmen/SE-lecture-2.pdf.

Chapter 6

Achieving Excellence in Software Development Using Metrics

We have seen metric development to suit the context, address needs and fulfill performance goals in Chapter 5. Now we shall discuss how to apply metrics to manage the software development cycle and achieve excellent results.

Let us look at a few representative metrics shown in the following paragraphs for developing the approach. For a comprehensive study of metrics, see Fenton [1], Grady [2], Stephan Kan [3], Pandian [4], and Putnam [5].

Examples of Project Metrics

Time to Deliver

> The first 90 percent of the code accounts for the first 90 percent of the development time ... The remaining 10 percent of the code accounts for the other 90 percent of the development time.
>
> **Tom Cargill**

This is the most serious concern of any project manager, time being a scarce resource that cannot be bought out but can only be saved. The trick is to track time performance using any convenient metric every milestone. The metric could be schedule variance or schedule slippage or earned value metrics (EVM).

Cost

This metric is tracked often as man days; the overhead can be added by a finance expert.

Quality

Every work product quality should be tracked by counting defects and normalizing the defect count by an appropriate expression of size.

Productivity

Size developed per man day is a common expression of this metric. There are several other definitions to choose from depending upon the purpose of measurement.

Time to Repair

The time taken to fix bugs is an important metric and is tracked automatically through the bug tracking tool.

Customer Satisfaction

This metric is obtained using quarterly surveys or annual surveys.

Requirement Volatility

Score creep or requirement volatility is a crucial metric and is to be closely tracked by the project team.

Examples of Product Metrics

Requirement Size

Requirement management suggests we measure the size in terms of number of features or number of pages or simply the number of requirements. Some use "use case points" or another specially designed metric.

Design Complexity

> With proper design, the features come cheaply. This approach is arduous, but continues to succeed.
>
> **Dennis Ritchie**

Function point (FP) and its variants can be used to judge design complexity; we need to measure the flow of information from and into every module. A summary of function point metrics is presented by Caper Jones [6].

A simple design metric has been proposed by Ball State University [7]:

The external design metric De is defined as

$$De = e1 \text{ (inflows * outflows)} + e2 \text{ (fan-in * fan-out)}$$

where inflows is the number of data entities passed to the module from superordinate or subordinate modules plus external entities, outflows is the number of data entities passed from the module to superordinate or subordinate modules plus external entities, fan-in and fan-out are the number of superordinate and subordinate modules, respectively, directly connected to the given module, and e1 and e2 are weighting factors.

The internal design metric Di is defined as

$$Di = i1 \text{ (CC)} + i2 \text{ (DSM)} + i3 \text{ (I/O)}$$

where CC, the Central Calls, is the number of procedure or function invocations, DSM, the Data Structure Manipulations, is the number of references to complex data types, which are data types that use indirect addressing, I/O, the Input/Output, is the number of external device accesses, and i1, i2, and i3 are weighting factors.

D(G) is a linear combination of the external design metric De and the internal design metric Di and has the form

$$D(G) = De + Di$$

The calculation of De is based on information available during architectural design, whereas Di is calculated when detailed design is completed.

The need for a Design Metric cannot be overemphasized. Good design makes the remaining phases of software development a smooth journey.

BOX 6.1 MEET THE EXPERT—ALLAN J. ALBRECHT: THE FATHER OF FUNCTION POINT

Allan J. Albrecht (1927–2010), the father of function points, never imagined that function points would be used by a large user community spread around several countries. Allan clearly outlined productivity as work product output divided by work effort. However, it was the development of the function point analysis concept as a means of identifying work product that has been his greatest contribution. Allan Albrecht's ideas shaped many careers.

Function point metrics were invented at IBM's White Plains development center in 1975. Function point metrics were placed in the public domain by IBM in 1978. Responsibility for function point counting rules soon transferred to the IFPUG.

The original formula was simple and can be used quickly without any hassle. The weights he used in the formula proved to be right and remained valid long after his invention. IBM advertised FP as a key to success and grabbed huge software development orders.

He developed the function point metric in response to a business need, to enable IBM customers for application software to state their requirements in a way that reflected the function of a proposed software system in terms that

> they could readily deal with, not the more technical language, e.g., SLOC counts, of the software developers. This was truly a major step forward in the state of the art, and in the state of practice of our profession.
>
> **John Gaffney**

FP could be applied to all programming languages and across all development phases; this helps development management and simplifies benchmarking. FP is a better predictor of defects than lines of code. FP is also a more effective estimator of cost and time.

Code Size

> A good way to stay flexible is to write less code.
>
> **Pragmatic Programmer**

The size of developed, deleted, modified, or reused code must be tracked as lines of code. Distinction can be made between executable and comment lines.

If function point metric is selected (in all phases), we can first calculate functional size using the original formula invented by Allan J. Albrecht:

EI	×	4	=	_____
EO	×	5	=	_____
EQ	×	4	=	_____
ILF	×	10	=	_____
EIF	×	7	=	_____
Total	FP		=	_____

where EI is the external input, EO is the external output, EI is the external inquiry, ILF is the internal logical file, and EIF is the external interface file.

The International Function Point User's Group (IFPUG) has introduced a detailed FP counting method in 1986 [8]. The results still agree with Albrecht's formula, but counting has become more precise. The IFPUG rules are time consuming to apply.

BOX 6.2 THE FULL FUNCTION POINT: A BREAKTHROUGH

The full function point (FFP) revolutionized size measurement. A new paradigm was invented: data movement is size. Full function points were proposed in 1997 with the aim of offering a functional size measure specifically adapted to real-time software. It has been proven that FFP can also capture the functional size of technical and system software and MIS software.

FFP distinguishes four types of data movement subprocess: entry, exit, read, and write, as identified in the context model of software. FFP makes use of the measurement principle: the functional size of software is directly proportional to the number of its data movement subprocesses.

Practice tends to show that the FFP approach, while offering results very similar to those of the IFPUG approach when applied to MIS software, offers more adequate results when applied to real-time, embedded, or technical software by virtue of the fact that (a) its measurement functions are not bounded by constants and (b) the level of granularity is more relevant to these types of software. Furthermore, in situations requiring the measurement of smaller pieces of software, the FFP approach offers a finer degree of granularity than the one offered by the IFPUG approach by virtue of the identification and measurement of subprocesses.

Serge Oligny and Alain Abran

Code Complexity

Again, function point is a good enough metric of code complexity. If a tool is available for measuring complexity, we can use the McCabe complexity number. This cyclomatic complexity measures the amount of decision logic in a single software module. Cyclomatic complexity is equal to the number of independent paths through the standard control flow graph model.

Highly complex modules are more prone to error, harder to understand, harder to test, and harder to modify. Limiting complexity helps. McCabe proposed 10 as the limit, but higher levels of complexity are in use.

Defect Density

A common metric to express quality is known as *defects per* kilo lines of code (*KLOC*) or *defects per FP*. The second formula can be used even in the design phase.

Defect Classification

Orthogonal defect classification or its variant is considered as a very critical defect measurement. From the classification, defect profile or defect signature can be extracted.

Reliability

> The price of reliability is the pursuit of the utmost simplicity. It is a price which the very rich may find hard to pay.
>
> **C.A.R. Hoare**

It is becoming a growing fashion to estimate reliability before dispatch. A suitable reliability model should be adopted for this purpose. A simple metric is failure intensity, meaning the number of defects uncovered per unit time of testing or usage.

Examples of Process Metrics

Review Effectiveness

This metric refers to the percentage of defects caught by review. More significantly, it draws our attention to the review process. It is now a well-established fact that review effectiveness improves quality and reduces time and cost, and this metric is worth watching.

Test Effectiveness

This a straightforward calculation of defects found by test cases and is a very good metric during the testing phase.

> Program testing can be used to show the presence of bugs, but never to show their absence!
>
> **Edsger Dijkstra**

Test Coverage

There are two expressions in use for this metric. First, we check how much of the requirement is covered by test cases (requirements coverage). Second, we track how much of the code is covered by testing (structural coverage) using a tool.

Subprocess Metrics

As the software development organization climbs the ladder of process maturity, visibility into what we do increases. One way of achieving this deeper visibility is to measure subprocesses.

For example, let us consider the case of "review process" and investigate construction of subprocess metrics. The process of review is measured by review effectiveness, the overall performance. To achieve subprocess measurement, review can be divided into the following subprocesses:

- Preparation
- Individual review
- Group review by meeting

Metrics can be installed to monitor these subprocesses, for instance, as follows:

- Preparation effort
- Individual review effort
- Individual review speed
- Group review effort
- Group review speed

Subprocess metrics provide the following attractive benefits

- We can build process performance models with the data.
- We can predict the overall process outcome.
- We can establish control at the subprocess level and increase certainty of achieving goals.

Achieving subprocess measurement is not easy. This requires voluntary effort from people, very similar to the case of data collection in the personal software process. We cannot force these metrics into the organization because these metrics "intrude" into creative efforts. Providing data at the subprocess level require great maturity and transparency. People resist subprocess data collection because of the following reasons:

- The reasons for subprocess data collection are not clearly known.
- People hate micromanagement.
- People are quick to realize that already collected data have not been used (a truth).
- People do not have time to think of process performance models.
- Scientific management is not a popular management style.
- People think that statistical process control is not relevant to software development.

Achieving subprocess metrics therefore requires a cultural transformation.

Converting Metrics into Business Information

Project Dashboard

Project metric data can be transferred to a project dashboard, preferably visual, as shown in Figure 6.1. The dash board must be updated, as completely as possible, after every milestone is performed. Some metrics such as customer satisfaction may be available less frequently, and this does not pose any serious problem. Achieving a milestone based dashboard to display metric results is the difficult first step.

BOX 6.3   THE RIGHT METRIC FOR PROJECT DELAY

Project delay is often measured using a classic metric "schedule variance." This is one of those variance metrics and has enjoyed the favor of many practitioners. The expression defining schedule variance is as follows:

$$\text{Schedule variance} = \frac{(\text{Actual schedule} - \text{Estimated schedule})}{\text{Estimated schedule}} \times 100$$

If a 6-month project is delayed by 5 days, the schedule variance, according to the previously mentioned definition, is calculated as follows: (5/180) × 100 = 2.8%. If the process specification limits are ±5%, this schedule performance is acceptable, or, to a cursory glance, the deviation is not alarming.

Instead of measuring delay as a normalized variance, measure a schedule slip expressed in actual calendar days by which the project is delayed, and a new meaning emerges. In projects that deal with the Y2K problem, even a single minute delay could play havoc. This is an example of the meaning of slippage in real time. In this project, process compliance to preset

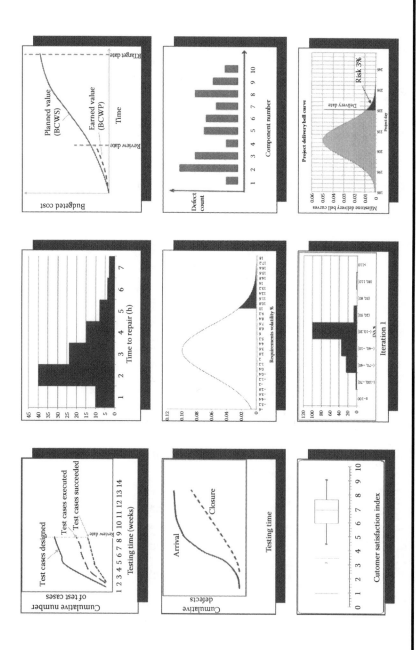

Figure 6.1   Project dashboard.

specification limits does not have meaning. Delay should be measured in real time. The cost of time depends on the context. For example, a single day delay in software delivery to large Aerospace Systems may incur huge costs. Where thousands of vendors are involved, scheduled delay can have catastrophic cumulative effects. Thus, the following metric must be used:

Schedule slip = (actual schedule − planned schedule) in days

The time loss must then be converted into monetary loss. Dollars lost due to schedule slip is a better metric than the percentage of process compliance. A 1-day slip might translate into millions of dollars in some large projects. Customers may levy sizable penalty on schedule slips.

Product Health Report

Product metric results can be organized in matrix format to present recorded information about each component. There are many ways one can organize this information. A simple form is shown as follows:

Metric	Component 1	Component 2	Component 3	Component 4	Component 5	Component 6	Component 7	Component 8	Component 9
Date tested									
FP									
KLOC									
Design complexity									
Code complexity									
Defect count									
Defects/FP									
Reliability									
Residual defects (predicted)									

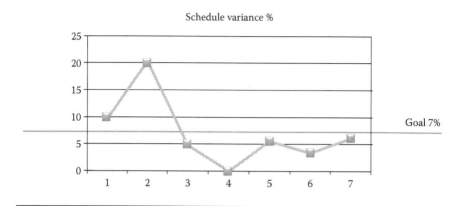

Schedule variance %

Goal 7%

Figure 6.2   Control chart on schedule variance.

Statistical Process Control Charts

From collected process data, we plot control charts. Simple statistical process control (SPC) techniques can be used to make process control a success. The control charts are sufficient to make project teams alert and deliver quality work products and simultaneously cut costs. Each chart can be plotted with results from completing various components, arranged in a time sequence, as shown in Figure 6.2. These charts are also known as *process performance baselines*.

Case Study: Early Size Measurements

Measuring size early in the life cycle adds great value. During the first few weeks of any development project, the early size indicators provide clarity to the project teams when things are otherwise fuzzy.

The Netherlands Software Metrics Users Association (NESMA) has developed early function point counting. According to NESMA,

> A detailed function point count is of course more accurate than an estimated or an indicative count; but it also costs more time and needs more detailed specifications. It's up to the project manager and the phase in the system life cycle as to which type of function point count can be used.
>
> In many applications an indicative function point count gives a surprisingly good estimate of the size of the application. Often it is relatively easy to carry out an indicative function point count, because a data model is available or can be made with little effort.

When use cases are known, by assigning different weight factors to different actor types, we can calculate use case point. This metric can be set up early in the project and used as an estimator of cost and time. Technical and environmental factors can be incorporated to enrich the use case point for estimation.

Similarly, when test cases are developed, a metric called *test case point* (TCP) can be developed by assigning complexity weights to the test cases. The sum of TCP can be used to estimate effort and schedule.

Alternatively, we can measure object points based on screens and reports in the software.

Use case points, test case points, and object points are variants of functional size. They can be converted into FP by using appropriate scale factors.

> *Measure functional size as the project starts.*
> *This will bring clarity into requirements and help in the estimation of cost, schedule and quality.*

Once functional size is measured, the information is used to estimate manpower and time required to execute the project. This is conveniently performed by applying any regression model that correlates size with effort. COCOMO is one such model, or one can use homegrown models for this purpose.

There is a simpler way to estimate effort. We can identify the type of software we have to develop. Yong Xia identifies five software types: end-user software (developed for the personal use of the developer), management information system, outsourced projects, system software, commercial software, and military software. On the basis of type, we can anticipate the FP per staff month, which can vary from 1000 to 35 [9]. Using this conversion factor, we can quickly arrive at the effort estimate. Once effort is known, we can derive time required, again by using available regression relationships.

Early metrics capture functional size and arrive at effort estimates; measurement and estimation are harmoniously blended.

Project Progress Using Earned Value Metrics

Tracking Progress

Whether one builds software or a skyscraper, earned value metrics can be used to advantage. To constrict earned value metric, we need to make two basic observations: schedule and cost are measured at every milestone.

The first achievement of EVM is in the way it distinguishes value from cost. Project earns value by doing work. Value is measured as follows:

> *Budgeted cost of work is its value.*

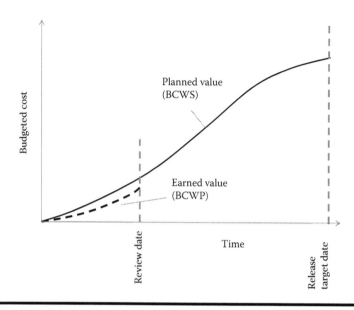

Figure 6.3   Tracking project progress.

In EVM terminology, this is referred to as *budgeted cost of work*. From a project plan, one can see how the project value increases with time. The project is said to "earn value" as work is completed. To measure progress, we define the following metrics.

Planned value = budgeted cost of work scheduled (BCWS)
Earned value = budgeted cost of work performed (BCWP)

An example of project progress tracked with these two metrics is shown in Figure 6.3.

At a glance, one can see how earned value trails behind planned value, graphically illustrating project progress. This is known as the *earned value graph*.

Tracking Project Cost

The actual cost expended to complete the work reported is measured as the actual cost of work performed (ACWP). Cost information can be included in the earned value graph, as shown in Figure 6.4.

In addition to the earned value graph, we can compute performance indicators such as project variances and performance indices. We can also predict the time and cost required to finish the project using linear extrapolation. These metrics are listed in Data 6.1.

An earned value report would typically include all the earned value metrics and present graphical and tabular views of project progress and project future.

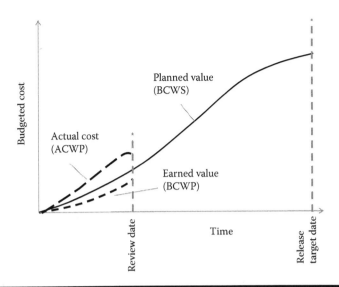

Figure 6.4   Tracking project cost.

Data 6.1   Earned Value Metrics

Core Metrics
Budgeted cost of work scheduled (BCWS) (also planned value [PV])
Budgeted cost of work performed (BCWP) (also earned value [EV])
Actual cost of work performed (ACWP) (also actual cost [AC])

Performance Metrics
Cost variance = PV − AC
Schedule variance = EV − PV
Cost performance index (CPI) = EV/AC
Schedule performance index (SPI) = EV/PV
Project performance index (PPI) = SPI × CPI
To complete schedule performance index (TCSPI)

Predictive Metrics
Budget at completion = BAC
Estimate to complete (ETC) = BAC − EV
Estimate at completion (EAC)
  EAC = AC + (BAC − EV)   Optimistic
  EAC = AC + (BAC − EV)/CPI   Most likely
  EAC = BAC/CPI   Most likely–simple (widely used)
  EAC = BAC/PPI   Pessimistic
Cost variance at completion (VAC) = BAC − EAC

The project management body of knowledge (PMBOK) treats earned value metrics as crucial information for the project manager. Several standards and guidelines are available to describe how EVM works. For example, the EVMS standard released by design of experiments (DOE) utilizes the EVMS information as an effective project management system to ensure successful project execution.

BOX 6.4   WHY IS SIZE METRIC MISSING?

We measure the size of what we build. If we construct a house, we measure the floor area. If we build a truck, we measure its length, height, and breadth, among other parameters. Size measurement is a hallmark of engineering.

When we build a software, we are reluctant to measure size. Some of us do not measure size at all. This is a disappointment for we have failed to uphold software engineering. Come to think of it, it is rather easy to measure size. If we find IFPUG function point or FFP time consuming and cumbersome, especially for large modules, we have alternatives to choose from. For example, we can use any of the following established metrics:

- Albrecht function point
- Mark II FP
- Object point
- Test case point
- Use case point
- PROBE
- Feature point

Doug Hubbard's book titled *How to Measure Anything* [10] can help to overcome mental barriers in measuring things considered impossible to measure. He observes that the perceived impossibility of measurement is an illusion caused by not understanding the concept of measurement, the object of measurement, and the methods of measurement.

You can build your own proprietary size metric. Some have constructed engineering size unit as a metric. Others have invented requirement size unit. Homegrown complexity metrics are doing the rounds as well. Having an approximate size metric is infinitely better than not having one.

 ## Review Questions

1. Mention your choice of a product metric in software development.
2. Mention your choice of a process metric in software development.
3. Mention your choice of a project metric in software development.
4. Mention your choice of a process for establishing subprocess metrics. Identify and list the subprocess metrics for the process you have selected.
5. Why is function point a success?

Exercises

1. Download a C program to convert Celsius to Fahrenheit from the web. Count size in lines of code.
2. For the previously mentioned script, count function point.
3. Count FFP for item 2. Compare FFP value with FP value.
4. Develop subprocess metrics for software design. Propose an equation for design complexity that makes use of the subprocess metrics.
5. The earned value metrics for a development project captured in the middle of the project life cycle are as follows

Earned value	1200 person days
Planned value	1300 person days
Actual cost	1500 person days

Calculate the following project performance indices:

Schedule performance index
Cost performance index

References

1. N. E. Fenton and S. L. Pfleeger, *Software Metrics: A Rigorous and Practical Approach*, International Thomson Computer Press, Boston, 1996.
2. R. B. Grady and D. L. Caswell, *Software Metrics: Establishing a Company-Wide Program*, PTR Prentice-Hall, 1987.
3. S. H. Kan, *Metrics and Models in Software Quality Engineering*, Addison-Wesley Longman Publishing Co., Inc., Boston, 2002.
4. C. Ravindranath Pandian, *Software Metrics: A Guide to Planning, Analysis, and Application*, CRC Press, Auerbach Publications, 2003.
5. L. H. Putnam and W. Myers, *Five Core Metrics: The Intelligence Behind Successful Software Management*, Dorset House Publishing, New York, 2003.
6. C. Jones, *Applied Software Measurement: Global Analysis of Productivity and Quality*, McGraw-Hill Osborne Media, New York, 2008.
7. D. Zage and W. Zage, *Validation of Design Metrics on a Telecommunication Application*, Available at http://http://patakino.web.elte.hu/SDL/tr171p.pdf.
8. R. R. Dumke, R. Braungarten, G. Büren, A. Abran and J. J. Cuadrado-Gallego, *Software Process and Product Measurement*, Springer, New York, 2008.
9. NESMA, *The Application of Function Point Analysis in the Early Phases of the Application Life Cycle. A Practical Manual: Version 2.0*, 2005. Available at http://www.nesma.nl /download/boeken_NESMA/N20_FPA_in_Early_Phases_%28v2.0%29.pdf.
10. D. W. Hubbard, *How to Measure Anything: Finding the Value of Intangibles in Business*, Wiley, New Jersey, 2010.

Suggested Readings

Aczel, A. D. and J. Sundara Pandian, *Complete Business Statistics*, McGraw-Hill, London, 2008.

Applying Earned Value Analysis to Your Project. Available at http://office.microsoft.com/en-in/project-help/applying-earned-value-analysis-to-your-project-HA001021179.aspx.

Astels, D. *Test-Driven Development: A Practical Guide*, Prentice Hall PTR, Upper Saddle River, NJ, 2003.

Cem, K. *Rethinking Software Metrics*, STQE, March/April 2000.

Estimated Function Point. Available at http://www.functionpoints.org/, http://www.functionpoints.org/resources.html, http://csse.usc.edu/csse/event/1998/COCOMO/18_Stensrud%20Estimating.pdf.

Function Point Estimation. Available at http://webcourse.cs.technion.ac.il/234270/Spring2013/ho/WCFiles/function%20points%20estimation.pdf.

Garmus, D. *An Introduction to Function Point Counting.* Available at http://www.compaid.com/caiinternet/ezine/garmus-functionpointintro.pdf.

Hunt, R. P., P. J. Solomon and D. Galorath, *Applying Earned Value Management to Software Intensive Programs.* Available at http://www.galorath.com/DirectContent/applying_earned_value_management_to_software_intensive_programs.pdf.

International Function Point Users Group, *IT Measurement: Practical Advice from the Experts*, Addison-Wesley, 2002.

Kitchenham, B. *Software Metrics—Measurement for Software Process Improvement*, NCC Blackwell, UK, 1996.

Kulik, P. *Software Metrics: State of the Art—2000*, KLCI, 2000.

Kusters, R., A. Cowderoy, F. Heemstra and E. van Veenendaal, editors, *Project Control for Software Quality*, Shaker Publishing, 1999.

McBreen, P. *Software Craftsmanship: The New Imperative*, Addison-Wesley Professional, 2002.

McCarthy, J. *Dynamics of Software Development*, Microsoft Press, 1995.

McConnell, S. *Professional Software Development*, Addison-Wesley, 2004.

Meli, R. and L. Santillo, *Function Point Estimation Methods: A Comparative Overview.* Available at http://citeseerx.ist.psu.edu/viewdoc/download?doi=10.1.1.33.6479&rep=rep1&type=pdf.

Norman, F. E., and M. Niel, *Software Metrics: Road Map*, CSD QM&WC, London, August 2000.

Parthasarathy, M. A. *Practical Software Estimation*, Addison-Wesley, 2007.

Platt, D. S. *Why Software Sucks—And What You Can Do about It*, Addison-Wesley Professional, 2007.

Putnam, L. H. and W. Myers, *Expert Metrics Views*, Cutter Information Group, 2000.

Wiegers, K. E. *Practical Project Initiation: A Handbook with Tools*, Microsoft Press, 2007.

Woodings, T. L. *A Taxonomy of Software Metrics*, Software Process Improvement Network (SPIN), Perth, 1995.

Chapter 7

Maintenance Metrics

Fusion of Frameworks in Software Maintenance

Software maintenance assimilates three management styles: project management, operations management, and service management. Maintenance metrics design mirrors this fusion. The big maintenance work gains from project management framework, the small tasks gain from operations management, and all maintenance tasks, by necessity, subscribe to service management framework. The project approach harmonizes all.

> Operations keeps the lights on, strategy provides a light at the end of the tunnel, but project management is the train engine that moves the organization forward.
>
> **Joy Gumz**

Maintenance engineering is based on multiple principles. On the one side, we have enhancement tasks (adaptive maintenance), which are mini projects to change the functionality of software [1]. Knowledge areas from Project Management Body of Knowledge (PMBOK) are often used as founding principles. On the other side, we have quick jobs of bug fixing (corrective maintenance), which are more like service tasks. Managing these tasks uses operations management, ITIL, ISO 20000, and CMMi SM concepts as founding principles. Metrics in these two different types of maintenance accordingly differ in their scope, nature, and intent. Metrics interpretations reflect the respective founding principles.

Occasionally, we also come across perfective maintenance where quality and performance of the software is improved. Perfective maintenance tasks are product and process improvement projects. IEEE standard defines perfective maintenance as "modification of a software product after delivery to improve performance or maintainability." It is now widely believed that software reliability enhancement is a social responsibility; it makes life safer. To fulfill these social expectations, responsible maintenance organizations gather such metrics. These are product and process performance metrics. Such metrics have rather complex definitions and are often based on mathematical models.

Corrective maintenance dominates the scenarios in some maintenance organizations. Some take up pure enhancements. In several other organizations, the three types coexist in a 3:1:1 ratio according to a study made by NC State University [2]. The ratio could be 1:3:1 in certain business contracts where the focus is on bug fixing. NC State University also mentions that corrective maintenance—the quick fixes—come in two styles: without document changes and with document changes in the 2:1 ratio.

In huge system enhancement contracts, such as in aerospace, a large number (thousands) of change requests are clubbed into a development package, and it goes through a full development life cycle for many years. Multiple variants of the product are released periodically. This becomes more complex than regular green field development projects. Additional activities such as impact analysis and regression testing are included to ensure that system integrity is maintained. For these projects, the metric approaches suggested in Chapter 6 are relevant and may be followed.

In maintenance projects, many data are collected by automated tools. However, the construction of metrics and models seems to be more difficult in maintenance projects than that in development projects. Maintenance tasks are shorter, and there is no time for manual data collection. Tool-collected data go into a database ("write-only" data base, Humphrey quipped) and is revisited by analysts who prepare reports for management and customers. Data, much less metrics, are not that visible to support teams.

Extraction of metrics from the database is performed based on the purpose at hand. The purpose could be weekly management reports or occasional construction of performance models.

Let us take a look at some typical maintenance metrics.

BOX 7.1   SHOULD MAINTENANCE TEAM MEASURE SOFTWARE RELIABILITY?

Reliability is a product metric and is often considered as a development metric. However, the role of reliability assurance shifts with time, from the development team to the maintenance team. Under maintenance tasks, reliability can either grow or deteriorate. Many support teams do not measure reliability unless the contract demands such a metric. As software evolves during maintenance, entropy sets in and quality gradually diminishes.

> As a system evolves its complexity increases unless work is done to maintain or reduce it; the quality of such systems will appear to be declining unless they are rigorously maintained and adapted to operational environment changes.
>
> **Lehman's Laws of Software Evolution**

It is becoming an implied need that the support teams look after software reliability too and hence must measure reliability.

Maintainability Index

A code is maintainable if it is understandable, modifiable, and testable, three factors identified by Syavasya [3]. A simple rule is

The more complex a code turns, the less maintainable it becomes.

A widely used and practical working metric, maintainability index, is defined as follows:

$$MI = 171 - 5.2\ln V - 0.23G - 16.2\ln LOC$$

where MI is the maintainability index, V is the Halstead volume, G is the cyclomatic complexity, and LOC is the count of source lines of code (SLOC).

Way back during coding, this metric could be used by a programmer to simplify a code in a systematic and measurable manner. The same metric can be used by the maintenance team to assess the application under maintenance. During the series of enhancements and feature additions, care may be taken to sustain high maintainability or even improve maintainability in preventive maintenance. This metric plays a fundamental role in the phases, coding, and maintenance:

What is measured, improves.

Tracking this metric, during evolution of software during maintenance, helps to regulate and guide maintenance efforts.

Change Requests Count

At the outset, the software size grows following an evolutionary path, release followed by release. Only a part of the software is not modified, as suggested by Lehman et al.

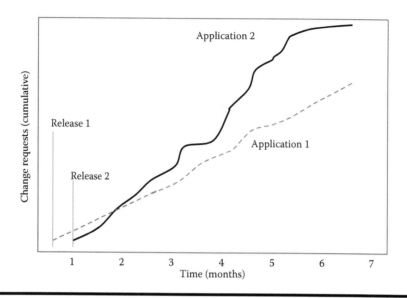

Figure 7.1 Change request (CR) arrival.

[4], and shows an example: the total size of a system in modules and the part of the system not touched at each release are plotted as a function of release number. Both measures are expressed as a percentage of the largest size achieved by the system.

In this dynamic environment, maintenance teams have to respond to change requests from the field—from customers. Change requests need to be validated and analyzed, and fixes must be designed, built, tested, and finally released. The arrival of change requests from two applications under maintenance is shown in Figure 7.1.

Change requests from application 1 show a trend of growth, suggesting more changes in the months to come, whereas change requests from application 2 seem to have reached a plateau region, suggesting negligible number of changes in the months to come. Cumulative plots of change request counts are of immense value to the maintenance team. They indicate work done and predict work to be done. The metric is a direct count and does not involve complex calculations. Meaningful are the patterns discernible to the human eye.

Customer Satisfaction Index

Measurement of customer satisfaction (CSAT) has a strong business purpose: it helps businesses grow. Maintenance metrics framework placed great emphasis on CSAT survey.

Customers take a more direct interest in bug fixing and tend to fill in customer satisfaction survey forms regularly and frequently. The factors selected for CSAT survey are unique and resemble typical sets used in the service industry. Zeithaml and Parasuraman's [5] RATER model measures the following factors:

Reliability
Assurance
Tangibles
Empathy
Responsiveness

More factors are added, and the previously mentioned factors are also tailored based on the business contract and customer's special requirements.

Customer satisfaction against each factor is measured using the Likert scale. However, a continuous scale from 0 to 10 is simpler, more granular, and has advantages, presented as follows:

Undesirable 0 o—o—o—o—o—o—o—o—o—o 10 Excellent

The maintenance CSAT factors are very different from the factors used in development projects, and the priorities are different. Preferably, the factors should be selected in consultation with the customer to have a perfect alignment with customer's expectations.

The CSAT index is a metric religiously collected in corrective maintenance operations; this is also the metrics more often seen by top management.

Resource Utilization

There is always a mismatch between the volume of maintenance tasks and the available resources. Workload is not constant but fluctuates. Projects could be understaffed or resources could be idle.

Typical maintenance projects are run with minimal resources to reduce overhead costs; as a consequence, there is a backlog. However, the resource utilization metric will be reported as 100% every month. This metric is a fallacy unless it is seen in the context of performance and customer satisfaction. If there is overtime and if it is also recorded, human resources utilization will touch 120% and more, as has been occasionally reported in the industry. Such high scores are alarming; employees could be put under stress, and quality of work may be compromised. Fatigue affects the way the team members communicate and empathize with customers. The healthy range of this metric is between 95% and 98%.

Along with this metric, team skill index can also be measured using the data collection form shown in Figure 7.2. When the team skill index is low, it has two significant consequences: customer satisfaction falls low, and it takes longer to fix bugs.

Service-Level Agreement Compliances

Service levels are specified in the maintenance contract. There are stringent specifications on the time to deliver each category of service. For example, the delivery time

Maintenance project ID
Team ref.
Assessment date
Evaluated by

| Team member | Skill score (scale: 0–10) | | | | | | |
| | Skill | | | | | | |
	Problem solving	Platform experience	Domain experience	Oral communication	Written communication	Analysis capability	Average
A							
B							
A							
B							
A							
B							
Team skill index (TSI) overall average							

If TSI >8, good
If TSI <8 but >5, poor, arrange training
If TSI <5, alarming problem, escalate it

Figure 7.2   Resource utilization—Team Skill Index (TSI).

could be 24 hours for express service, 48 hours for next priority tasks, a week for regular tasks, a fortnight for odd tasks, and so on. There are different performance slabs for different task categories. If delivery had been made within the stipulated time, the service-level agreement (SLA) is complied; if not, it is a noncompliance and a breach of contract. The customer virtually controls the maintenance team by closely monitoring the SLAs. The criteria are often designed to minimize risk to the customer.

The SLA compliance metric is based on counts, defined as follows:

$$\text{SLA compliance} = \frac{\text{Number of deliveries that met SLA}}{\text{Total number of deliveries}} \times 100$$

Customers sign different SLAs with the maintenance organization for each of the delivery attributes such as quality, response time, priority levels, and volume delivered per month. Every month, the percentage of the SLA compliance level is measured for each. Noncompliance may attract penalties, and hence these metrics are respected and sincerely tracked.

SLA metrics enable teams to perform within the limits set. They do not measure the exact performance but only register whether the SLA is met or not. For example, while providing a work around governed by a 48-hour SLA criterion, the agent does not report the actual time. Even if the agent completes the job within 12 hours, delivery will be officially logged only at the 48th hour because the SLA says so. Twelve hours is unaccounted for. This is where Parkinson's law plays a role:

> *Work expands to fill the time available.*

There is no motivation to do your best, but to do just so much, the bare minimum, merely to avoid penalty. Under SLA, we never know the true capabilities of teams.

SLA compliance is a business metric in its strict sense.

Percentage of On-Time Delivery

Of all the service attributes, time is the most crucial. A special metric is constructed to track the % of deliveries made on time. The on-time delivery (OTD) metric elicits respect from the maintenance team because of its inherent business context.

This metric is very different from schedule variance (SV), which is a process metric and measures delay. As a process metric, even the magnitude of delay is captured as information. In OTD, the magnitude of delay is not captured. ITD is a discrete metric, whereas SV is a continuous metric. OTD captures partial information, whereas SV captures complete information. If we have not delivered on time, there is a consolation we get in measuring the delay. Even if the delay is small, no mercy is shown. The delivery is said to have failed. OTD belongs to a pass/fail world of hard decisions.

Enhancement Size

The enhancement size metric helps in understanding the enhanced job better, besides serving as an estimator of cost, schedule, and quality. The Netherlands Software Metrics Users Association (NESMA) guideline "Function Point Analysis for Software Enhancement Version 2.2.1" defines enhancement as changes to the functionality of an information system, so-called adaptive maintenance. Enhancement involves three possible tasks:

- Addition of functionality
- Deletion of functionality
- Change of functionality

Addition of functionality is measured as added FP. The deletion of functionality is measured as 0.40 × deleted FP. Changed functionality is measured as impact factor × changed FP; the impact factor could take values from 0 to 1.

The total is called the *enhancement size*, calculated as follows: enhancement function point (EFP) = added FP + 0.40 deleted FP + IF × changed FP.

The previously mentioned equation shows the effect of deletion and change. To work with EFP is a good practice.

As a proxy, LOC can also be used to judge enhancement size, though such a metric may not be available early in the enhancement life cycle and may not help in estimation. It is a pity to see that some projects do not include deleted size and changed size in their calculation.

BOX 7.2   BUG REPAIR TIME METRIC

A maintenance manager desired to statistically establish the team's bug repair capability and circulated a form to gather data. Senior engineers responded truthfully. The time spent on bug fixing was approximately 5–6 hours every day. A new recruit did not know how to respond and simply wrote "From 9 am till 6 pm, I spent time on bug fixing"; a statement too good to be true. The new recruit filled in what he thought as an appropriate value and not what he actually did. Bug repair time is a metric difficult to collect and even more difficult to validate.

Bug Complexity

In a lower scale of measurement, bug complexity is measured as high, medium, or low by the maintenance engineer. This is a subjective measurement, but it works.

Bug complexity can also be assessed on a continuous scale of 1–10, 10 being the most complex and 1 being the least. This is still subjective but has better granularity and has the extra advantage of being a numerical expression allowing further calculations.

The objective treatment of bug size considers factors driving bug fixing effort. In one model by Andrea De Lucia [6], the number of tasks required and the application size are considered as the factors, and the linear regression equation relating these two to effort is constructed.

The purpose of measuring bug complexity is to use the answer to predict bug fixing effort. That means the purpose is to build an estimation model. During analysis, such an estimate is made by the maintenance engineer, usually on the fly. Because analysis of the bug is made to fix the bug and not to build a model, exploring additional factors or increasing the depth of measurements is not suggested. Bug fixing is the main objective, model building a concomitant one. Moreover, we do not need extraordinary precision in estimating the bug fixing effort; we need a reasonably useful indicator.

Do not measure with a micrometer, mark with a chalk and cut with an axe.

Murphy's Law of Measurement

There are several equally simple ways of building an estimation model, including estimation by analogy and proxy-based estimation.

Bug complexity is thus measured with consciously selected level of approximation. Later, this is going to affect the effort variance metric.

Effort Variance (EV)

The formula for the effort variance metric remains the same as before:

$$\text{Effort variance} = \frac{(\text{Actual effort} - \text{Estimated effort})}{\text{Estimated effort}} \times 100$$

However, context and interpretation change in maintenance. This metric can be easily applied to enhancement projects where estimation is performed reasonably well, and the metric accordingly carries full meaning. It is beneficial to calculate effort variance twice, first with the initial estimate and later with a revised estimate after the change request is better understood. Even after the second estimate, the first estimate is still used as a budget control metric and the second as a process control metric.

In bug fixing, effort variance can be calculated, if at all, only approximately because of approximations in estimation.

Schedule Variance (SV)

The SV metric is treated like effort variance. Often, this is restricted to enhancement projects and not implemented in bug fixing tasks for two principal reasons: bug fixing is tightly controlled by SLAs, and the actual time of fixing is not available. Many times, bug fixing happens without estimation.

Quality

Quality of Enhancement

Quality metric is calculated by dividing defect count by size and is expressed as defects per EFP. The quality of each release is monitored.

Quality of Bug Fix

Sometimes, maintenance activities inadvertently harm quality; while fixing a bug, another could be introduced. In a typical bug-fixing environment, the support team does not know if a fixed bug opens in the field. If the same bug returns, people still may not detect this arrival because there is no traceability. Usage-triggered failures seldom come to the knowledge of the bug fixer. Bug arrival rate is not usually connected with quality because no one connects the dots. In such a fluid situation, unless quality metric is defined and collected, the quality of the software under maintenance cannot be known and improved. However, such a step needs to be negotiated with the customer and be seen as a business need. If quality improves, maintenance cost will come down and the customer will benefit; but preventive maintenance has to be paid.

Productivity

Productivity can be expressed in different ways. We suggest the EFP metric per man month. This metric eventually controls cost of maintenance; it helps in cost control.

For measuring bug fixing productivity, the number of fixes per man month could be a basic metric.

Measuring productivity is straightforward, but estimating productivity from contributing factors is not. If the metric includes estimation (which in a broader sense is a right expectation), then making use of models such as COCOMO may be included in our purview.

Time to Repair (TTR)

This is a metric automatically collected by the bug-tracking tool. If one wants to improve the process of bug fixing, optimize it, and achieve excellence, subprocess metrics such as (1) time for replication, (2) time for analysis and design, and (3) time for implementation and testing can be collected. These subprocess metrics can be obtained by a quick survey. The bug tracker tool may not be equipped to collect subprocess metrics. Such surveys are performed occasionally to obtain information to improve the process. It is quite possible to collect metrics at more granular levels, measure analysis time separately, and design time separately if the cost is justified by expected gain. One can go a step further and apply lean techniques such as value stream map analysis, waiting time analysis, and idle time analysis. This metric will help to make the operation more efficient.

Most certainly, one does not choose subprocess metric for regular data collection till the organization achieves high maturity and people volunteer to provide "personal" data. However, whatever be the level of granularity, bug repair time is one of the most effective and beautiful metrics in software engineering.

BOX 7.3   THE QUEUE

Bugs form a queue, and customers wait for fixes. Customer satisfaction increases with response time and quality; both depend on human resources. When resource utilization is 100%, customer satisfaction is less than the best. When resource utilization is 90%, customer satisfaction improves. Moral of the story: some human reserves must be maintained to boost customer satisfaction, and there is a trade-off between the two. Maintenance organization keeps buffer resources and operates at less than 100% resources utilization because losing customer satisfaction is costlier.

Backlog Index

This index measures the percentage of bugs in the backlog queue and represents an operational challenge in maintenance projects.

Bug Classification

Classification is categorical data. To understand bugs better, they are classified in the bug tracker database. Typical fields are origin, cause, trigger, type, and solution. More can be added; ODC use may be considered. A periodical analysis of bug distribution among the various classifications is a good revelation of the problem.

Fix Quality

Bug fixes could have problems. Bugs may reopen and cause rework and delay. The percentage of bug fixes that are performed right the first time can be measured. Rework can be measured for cost control.

Refactoring Metrics

To improve quality of software and reduce maintenance costs by refactoring, we measure coupling and cyclamate complexity. Coupling measures the flow of data between modules. Cyclomatic complexity measures complexity in module structure.

Reliability

Reliability of the product under maintenance can be computed from "failure intensity," which is the number of bugs reported per month. From this metric, we can judge whether the product is recovering or crashing. A time series plot of cumulative bugs discovered called *reliability growth curve* can be used as a visual representation of reliability. The mean time between failure also can be computed.

Metric-Based Dashboards

We would consider the top seven metrics for constructing a graphical dashboard for managing software maintenance. The other metric data may be presented in the form of a tabular record. The choice of metrics for graphical presentation is entirely up to the maintenance teams. Stark uses the following metrics [6]:

1. Backlog
2. Cycle time
3. Reliability
4. Cost per delivery

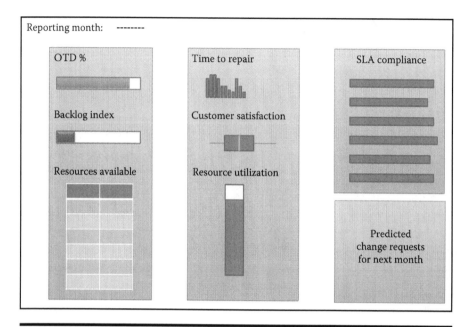

Figure 7.3 Maintenance dashboard.

5. Cost per activity
6. Number of changes by type
7. Staff days per change
8. Percentage of invalid change requests
9. Complexity assessment
10. Maintainability
11. Computer resource utilization
12. Percentage of content changes by delivery
13. Percentage of OTD

Figure 7.3 shows an example of a dashboard built with seven metrics.

BOX 7.4 LEHMAN'S LAWS OF SOFTWARE EVOLUTION

The laws of software evolution refer to a series of laws that Lehman and Belady formulated in 1974:

1. Continuing change—a system must be continually adapted or it becomes progressively less satisfactory.
2. Increasing complexity—a system evolves as its complexity increases, unless work is done to maintain or reduce it.

3. Self-regulation—a system evolution process is self-regulating with distribution of product and process measures close to normal.
4. Conservation of organizational stability (invariant work rate)—the average effective global activity rate in a system is invariant over product lifetime.
5. Conservation of familiarity—a system evolves all associated with it; developers, sales personnel, and users, for example, must maintain mastery of its content and behavior to achieve satisfactory evolution. Excessive growth diminishes that mastery. Hence, the average incremental growth remains invariant as the system evolves.
6. Continuing growth—the functional content of a system must be continually increased to maintain user satisfaction over its lifetime.
7. Declining quality—the quality of a system will appear to be declining unless it is rigorously maintained and adapted to operational environment changes.
8. Feedback system evolution processes constitute multilevel, multiloop, and multiagent feedback systems and must be treated as such to achieve significant improvement over any reasonable base.

Review Questions

1. What is the most important metric in adaptive (enhancement) maintenance?
2. What is the most important metric in corrective maintenance?
3. What is the most important metric in perfective maintenance?
4. Which among the previously mentioned three metrics is the toughest to collect?
5. If you were to design a dashboard for a support project and if you were asked to limit the number of metrics to just seven, which seven would you choose?

Exercises

1. Develop a metrics list for a support project if the adaptive–corrective–perfective maintenance task ratio is 3:10:2.
2. Develop a metrics list for an exclusive support contract for the perfective maintenance of software with an express goal of improving maintainability.

References

1. NESMA, *Function Point Analysis for Software Maintenance Guidelines, Version 2.2.1*, Professional Guide of the Netherlands Software Metrics Users Association, Netherland 2009.

2. NC State University Communication. Available at http://agile.csc.ncsu.edu/SEMaterials /MaintenanceRefactoring.pdf.

3. C. V. S. R. Syavasya, An approach to find out at which maintainability metric, software maintenance cost is less, *International Journal of Computer Science and Information Technology*, 3(4), 4599–4602, 2012, ISSN:0975–9646.

4. *Journal of Software Maintenance and Evolution: Research and Practice*, John Wiley & Sons, Ltd.

5. V. Zeithaml, A. Parasuraman and L. Berry, *Delivering Quality Service*, Simon and Schuster, 1990.

6. G. E. Stark, *Measurements to Manage Software Maintenance*, Technical Note, The MITRE Corporation, Colorado Springs, CO, July 1979.

Suggested Readings

Aczel, A. D. and J. Sounderpandian, *Complete Business Statistics*, McGraw-Hill, London, 2008.

Available at http://www.refactoring.com/.

Available at http://wiki.java.net/bin/view/People/SmellsToRefactorings.

Chapter 8

Software Test Metrics

The broad benefits of metrics discussed in Chapters 1 and 5 are relevant to software testing too. Although testing is part of the full life cycle, it has become a project of its own kind, with its own unique objectives. Testing is conducted to find defects and to improve product quality. The three objectives of any project, that is, faster, better, and cheaper, also apply to testing. Metrics would help to push testing to greater levels.

Project Metrics

Definitions of test project metrics are exactly same as metrics definitions in any project, with the same meaning and purpose. Project metrics help to conserve project resources and make optimal use of them. Project metrics track requirement changes and help to take corrective measures when requirement changes threaten the project. Project metrics also help to satisfy customers.

Schedule Variance

In testing projects, delivering on time is important. Measuring time and schedule variance as the testing milestones are crossed would give a feedback to the team to work toward meeting the delivery schedule.

Effort Variance

Completing testing within budgeted effort is the next concern. An effort variance metric would help to control the cost of testing.

Cost

The cost of testing per release is measured. Cost variance tracks dollars spent in excess of the budget. In addition to human cost, we need to consider investment on tools and outsourcing and see if we can execute testing within budget.

BOX 8.1   S CURVES IN TESTING

There are a few S curves used in testing. Cumulative defects arrived is the first curve. This is also called the reliability growth curve. This curve ends in a plateau zone beyond which further testing does not find defects. The product is said to have become stable, as far as we know, and can be shipped. Cumulative test cases executed is another S curve. The pattern of this S curve tells a lot about the nature and quality of test progress. Experienced testers can interpret this pattern and take appropriate decisions.

Human Productivity

From the view of project management, we look at defects found per tester. The result is used to provide feedback to testers and hence improve test results. Defect discovery rate is one of the metrics used as an index of team productivity.

Requirement Stability

Testing closely follows requirements. Test cases mirror use cases. Hence, the biggest uncertainty in testing is requirement stability. This is measured and tracked. The requirement stability index (RSI), also called requirement volatility, is defined as follows:

$$\text{RSI} = \frac{\text{Original req} + \text{Req changed} + \text{Req added} + \text{Req deleted}}{\text{Original req}}$$

RSI is a metric that might already exist in the metrics system developed for the entire development project. One has just to reuse it.

Resource Utilization

From an operational perspective, resource utilization is a key metric. This metric is extended to tools, systems, and people.

Customer Satisfaction

Eventually, testing is a service. Customer satisfaction must be measured by conducting surveys to improve service quality.

Test Effectiveness

This metric captures the percentage of defects found by testing. Of course, the stretch goal of the test team is to reach 100% effectiveness. The metric is defined as follows:

$$\text{Test effectiveness} = \frac{\text{Defects found by tests}}{\text{Defects found by tests} + \text{Defects found by business users}}$$

BOX 8.2   MEASURING THE RETURN ON INVESTMENT OF TEST AUTOMATION

It is good to automate test cases. Test automation needs creativity. Carefully directing test automation will result in cost saving. To make sure that automation is kept profitable, we introduce a metric ROI of test automation.

Regression tests can be automated; the ROI is great. When manual testing is difficult or impossible, automation is required. The simulation of a test scenario is difficult manually and is best performed through automation. In special cases such as testing a firmware, automation is the only way. In testing middle layers with missing upper or lower layers, automation is the only way.

Investment on automation may yield benefits beyond the current project. The tool must be generic enough to accommodate the needs of different projects. The tool need not aim at solving the requirements of a single project; it must be planned to address the needs of upcoming projects. It is an organizational asset.

ROI from automation may vary from two to ten, typically.

Process Metrics

The testing process is managed with several metrics, continuously tracked during testing life cycle. A few of them are cumulatively graphed to derive deeper meanings.

Defect Removal Efficiency

This metric is used in a special context in testing projects. There are other definitions for this metric in other contexts. Defect removal efficiency (DRE) in testing means the number defects removed per unit time, defined as follows:

$$DRE = \frac{\text{Number of defects}}{\text{Detection time} + \text{Resolution time} + \text{Retesting time}}$$

The inverse of this number is called defect turnaround time (TAT), defined as follows:

$$TAT = \frac{1}{DRE}$$

Test Cases Count

We can cumulatively count test cases designed, executed, and succeeded until any point of time chosen for inquiry. This count makes more meaning if plotted as a cumulative chart as shown in Figure 8.1. These charts measure dynamic changes in test case counts.

Discernible in these charts are a few useful metrics: the percentage of successful test cases and the percentage of executed test cases. These two metrics provide

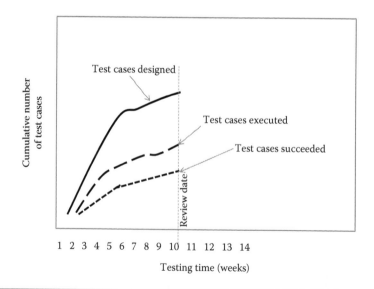

Figure 8.1   Cumulative test cases count.

useful feedback to the test management. The test team would strive to fill in the gaps in these areas. The first refers to quality test cases. The second pertains to a commitment to execute test case.

Test Coverage

Functionality Coverage

It is vital to know the proportion of requirements (functionalities) covered by test cases. This is the simple and most widely used test coverage metric. This metric helps to control and improve coverage and makes the application more usable by customers. As testing is in progress, coverage will increase in the light of this metric and will reach 100% in the ideal case.

Code Coverage

Yet another coverage metric traces the proportion of code covered by test cases.

Coverage helps eliminate gaps in a test suite.

This is a tenuous metric. Higher coverage does not mean assurance of better quality. An experienced tester knows to take a balanced view on this and make sure a minimum coverage has been achieved and critical paths have been covered.

Code coverage is a very useful metric.
However, you need to know how to use it.

Coverage metric tools are available to track line, statement, block, decision, path, and condition coverage. They provide excellent reports with back tracing and help achieve higher test efficiency.

BOX 8.3   UNIT TEST DEFECT DATA

A unit test is cost effective. It improves reliability. It reveals bugs that are otherwise devious. A unit test needs design knowledge and is best performed by testers with design knowledge. For best results, testers can collaborate with designers and developers. The level of thoroughness and documentation depends on test strategy and goals. However, often enough, unit tests are not well documented, and unit test defects are not entered into the bug tracker. There is not much to motivate the designer except project objectives and leadership drive.

> Program testing can be used to show the presence of bugs, but never their absence.
>
> **Dijkstra**

Some organizations have achieved partial success in getting unit test defect data. At the least, they collect defect count per module. The defect count is only partial because not all defects are counted. Even partial defect data would help in understanding defects in the product before even the formal test cycles start.

Percentage of Bad Fix

This is a subprocess metric, governing the quality of defect resolution. Bad fixes contains technical risk in testing, besides fuelling rework cost. In addition to poor resolutions, fixing an existing bug can sometimes create additional bugs or expose other bugs.

Product Metrics

The ultimate purpose of testing is to evaluate and measure the quality of the software under test (SUT), hence the importance of product metrics in testing.

Defect Counts

Defect Arrival Rate

The number of defects detected per day is the defect arrival rate. The cumulative number of defects found can be plotted as a pattern.

Defect Closure Rate

The number of defects closed per day is the defect closure rate. The cumulative plot of this number is the defect closure pattern. Figure 8.2 shows the closure pattern together with the arrival pattern. The gap between the two patterns has a self-evident meaning. The figure shows the nearly completed detection process while closure seems to be ongoing.

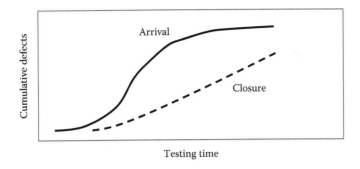

Figure 8.2 Defect arrival closure patterns.

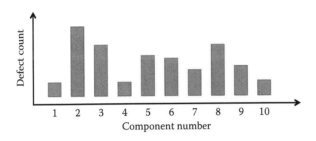

Figure 8.3 Defect per component.

Component Defect Count

Defects found in individual components can be separately analyzed. Figure 8.3 shows a bar graph of defects found in each component. This figure reveals defect proneness of some components.

Component Defect Density

When defect count is normalized (divided) by size, we can calculate the metric defects/function point (FP); we obtain fresh information about the quality levels of each component. Now there is an objective basis to do an intercomparison of quality of components. These data can be obtained at the first round of tests and updated during subsequent test iterations. The first round of results are a fairly good indicator of the quality levels. This will serve as a guide in developing test plans for the remaining rounds, and help capture more defects.

> **BOX 8.4 SMOKE TEST DEFECT DENSITY**
>
> Smoke tests are very useful; they shorten the testing cycle. They ensure that the code is working and the build is stable. They should be capable of exposing major problems. It is instructive to calculate defect density for each component with smoke test defects. Smoke test defect density is an indicator of risks in the components and an early predictor of problems. The real power of metrics is realized when metrics are used to predict. The smoke test is simple, fast, and cheap, but smoke test defect density is priceless.
>
> Smoke tests must be identified during requirements.
>
> For example, safety requirements can be smoke tested very early in the project.
>
> The smoke test sets the minimum entry criteria for starting test execution.

Defect Classification

During defect capturing, we can register the different attributes of defects: defect type, defect severity, phase injected, trigger, cause, and complexity. Then we can classify defects according to each of the attributes and derive broad perspectives about defect population. Classification is also a measurement. The defect attribute analysis report is a valuable input to the development team to support defect prevention.

Testing Size: Test Case Point

The test case point (TCP) can be used as an estimator for the entire software development project also.

Testing size can be measured in terms of TCP. Each test case is assigned a weight based on test case complexity and test case type. Test case complexity depends on test data, checkpoints, and precondition. The weights are as follows:

Checkpoint 1
Precondition 1, 3, 5
Test data 1, 3, 6

The rules for assigning weights are provided by Nguyen et al. [1]. Unadjusted text case points are computed with the above weights. Adjustment is performed according to the test case type. In a case study, the author reports, "The number of estimated TCPs was 8783 for the total 1631 test cases of the system, an average of 5.4 TCPs per test case." The total TCP represents testing size and can be used to estimate effort and time required to complete testing. Also in an experiment, Nguyen et al. noted that testers spent, on average, 3.21 minutes to execute each TCP.

Patel et al. [2] have extended TCP calculation. They have determined TCP separately for test case generation, manual execution, and automated execution. After these calculations, they add all the TCPs to determine the total TCP for testing.

Risk Metric

Testing risk is measured using the FMEA table and expressed as risk priority number (RPN).

Jha and Kaner [3] have applied FMEA to detect risks and prioritize them for mobile application.

Failure mode is the way a program could fail.

They have prepared a catalogue of failure modes under different quality criteria. The catalogue is pretty detailed and can be a useful reference to testers. These failure modes catalogued by Jha and Kaner are domain specific; testers have to identify their failure modes in the software testing, by using experience and knowledge.

Experienced testers know failure modes in software.

Having identified the failure modes, then they are prioritized using RPN. Then test cases are first developed to cover top 20% risks. Such inputs help testers to plan testing using the paradigm "risk-based testing." This enables testers to begin with a small set, learn from the results, and progress. Kaner [4] advocate risk-based test prioritization, test lobbying, and test design.

For every feature, we can measure RPN and prioritize features for testing. For every product, we can measure RPN and prioritize areas for testing.

Allocate more resource and earlier testing for bigger numbered (RPN) areas.

Cem Kaner

The purpose of risk measurement is to prioritize, to test less, to test sooner, and to obtain more knowledge.

Predicting Quality

As testing progresses, the quality of SUT would be progressively measured. Halfway through in testing, we have sufficient data to predict the quality level of the product at the end of completing test execution. We should be in a position to predict residual defects when testing is performed and the application is released. On the basis of this prediction, we can make a decision to stop testing once the targeted quality level is achieved. If a quality level is defined as 99.999% correct code and if prediction shows only 0.001% defects are undiscovered, we can release the application.

In this context, the purpose all test metrics is to make this prediction. Chapter 21 describes a method for predicting defects.

Metrics for Test Automation

Test automation requires special effort, almost a project on its own merit. Automation promises to rapidly find defects that manual testing cannot find. This increases the quality of SUT.

Automation could be a huge effort, especially for large projects.

Windows NT 4 had 6 million lines of code, and 12 million lines of test code.

By using test automation metrics, this huge effort could be managed better and more defects could be found in the SUT.

Return on Investment

Return on investment (ROI) of automation is a strategic measure that justifies automation. The simple formula for ROI is a direct ratio given as follows:

$$\text{ROI} = \frac{\text{Cost saved by automation}}{\text{Cost of automation}}$$

Cost savings from automation is usually distributed across several projects that benefit from the automation scripts; calculating ROI from single project information is not fair.

Percentage Automatable

Another lead metric in test automation is % Automatable. This metric must be established upfront:

$$\% \text{ Automatable} = \frac{\text{Number of test cases that can be automated}}{\text{Total number of test cases}} \times 100$$

Deciding on the test cases to be automated is a critical decision. Not every test case could be automated. Some defects could be found only by slow manual testing. Although speed is an advantage, only some test cases should be automated.

Automation Progress

To track the automation project, we can measure automation progress by setting up the following metric:

$$\text{Automation progress} = \frac{\text{Number of test cases actually automated}}{\text{Test cases selected for automation}} \times 100$$

Case Study: Defect Age Data

Defect age is a very well-known metric in the industry, but only a limited number of organizations collect this metric. The time defect spends in the product, from injection till resolution, is defect age. This can be calculated from the defect data in the bug tracker if appropriate fields have been set and populated. The fields required are phase injected and phase detected. It is now evident that the defect database is maintained throughout the project life cycle. Moreover, the information in the column phase injected is only obtained by reasoning during causal analysis and not from the log book. The term data in testing includes even reasoned, intuitive judgments.

Defect age = 1 + phase detected − phase injected

Some people use the formula without the "1"; they just recognize the incremental difference and choose to keep the base value at 0. If a defect is detected in the same phase where it was injected, defect age is 1. If discovery happens in the next phase, defect age is 2. If a defect is injected during the requirement phase but discovered in the test phase, defect age is 4.

This case study is about a testing project with a defect database with 32 fields and quite rich in raw data. However, no one derived the metric defect age although the possibility was there.

Much after the release, data mining by a QA analyst revealed the defect age metric. Motivation for this metric creation from available data had to come from an unexpected direction. Managers asked for a model for defect economics, and one of the factors driving cost was obviously defect age. Supporting metrics in this model building were the cost of finding defects and the cost of fixing defects. Both could be calculated from the data available in the database. The cost of finding defects was labeled as part of the cost of appraisal, and the cost of fixing defects was labeled as rework or the cost of poor quality as unmistakable influences from the cost of quality framework. However, these metrics were reported as part of defect age study and not COQ study. Life of bugs was nearer to people than formal frameworks. Testers perceived COQ as a larger metric for senior managers.

(Elsewhere, the cost of quality is also computed within the testing process. Three components of COQ are used: prevention, appraisal, and failure. Failure means rework on test cases and retesting, appraisal refers to reviews, and prevention includes training and defect prevention effort.)

The goal of creating the defect age metric was to establish a cost model by discovering a relationship between defect age and cost of defects.

> The most important quality metric is cost of failure.
>
> **Crosby**

Designers did not want to think of the cost of defect as the cost of failure, but testers did. Despite a controversy, the defect age metric became a potential metric. There was an engineering concept wherein defect age reflects product reliability. The lesser the defect age, the more the reliability. The project team strived to reduce defect age.

Another purpose of creating the defect age metric was to check the 1:10:100 rule of the cost of fixing defects. The rule says if the early fixing of defects costs a dollar, late fixing would attract exponentially increasing costs. Deep set defects are difficult to find and costly to fix. Data revealed that in that project, the rule was 1:2:4.2. There was no dramatic rise of cost when defect age increased.

Review Questions

1. How many metrics would you use in a test dashboard in a testing project?
2. What are the metrics that can be used in unit testing?
3. What are the metrics that can be used in smoke tests?
4. Suggest simple ways of assessing reliability of software before release.
5. What metric data will be used while making a decision about stopping testing?
6. Compare test effectiveness with test efficiency.
7. Mention the names of two commonly used S curves in testing.
8. Relate defect age to the cost of testing. In your opinion, what would be the expression for such a relationship?

Exercises

1. Develop a minimum set of metrics you would maintain for a testing agency who provides testing services to software developing companies.
2. Develop a risk metrics system for risk-based testing in a software development project.
3. Suggest a matrix format for checking requirement traceability with test cases.
4. Develop a metric system to be used for managing test automation.
5. Develop a template for a one monthly test report based on metric data. Mention the names of metrics and the charts you would use.

References

1. V. Nguyen, V. Pham and V. Lam, *Test Case Point Analysis: An Approach to Estimating Software Testing Size*. http://csse.usc.edu/csse/TECHRPTS/2007/usc-csse-2007-737/usc-csse-2007-737.pdf.
2. N. Patel, M. Govindrajan, S. Maharana and S. Ramdas, *Test Case Point Analysis*, White Paper Version 1.0, Cognizant Technology Solutions, April 11, 2001.
3. K. A. Jha and J. D. Cem Kaner, *Bugs in the Brave New Unwired World: A Failure Mode Catalog of Risks and Bugs in Mobile Applications* (personal communication). Copyright © 2003.
4. C. Kaner, Risk-based testing: Some basic concepts, *QA Managers Workshop*, April 2008.

Suggested Readings

Kaur, A., B. Suri and A. Sharma, Software testing product metrics—A survey, *Proceedings of National Conference on Challenges & Opportunities in Information Technology (COIT-2007)*, RIMT-IET, Mandi Gobindgarh, March 23, 2007.

Kelly, D. P. and R. S. Oshana, Improving software quality using statistical testing techniques, *Information and Software Technology*, 42, 801–807, 2000.

Nirpal, P. B. and K. V. Kale, A brief overview of software testing metrics, *International Journal on Computer Science and Engineering*, 3(1), 204–211, 2011.

Poore, J. H. and C. J. Trammell, *Application of Statistical Science to Testing and Evaluating Software Intensive Systems*. Available at http://sqrl.eecs.utk.edu/papers/199905_cjt_hmc.pdf.

Prabhakar, J. Test execution through metrics, *STEP-AUTO Conference*, Bangalore, Test Management, September 19, 2007.

Chapter 9

Agile Metrics

The purpose and character of agile metrics empathize with the Agile Manifesto.

> Individuals and interactions over processes and tools.
>
> **Agile Manifesto**

The proper design and implementation of agile metrics can add value and enable the agile way of developing software.

BOX 9.1   ANALOGY: DECIBELS

The human ear transforms sound waves into audio signals. When the sound wave intensity increases tenfold in strength, the response of the human ear goes up 1 point. When sound intensity increases a hundred times, the human ear registers strength of 2 points. The response is logarithmic. Sound level is measured in decibels, which are essentially logarithms of sound wave intensity.

Sound Intensity	Relative Response of Ear
10	1
100	2
1000	3
10,000	4
100,000	5

The ear can process and respond to a remarkably wide range of sound intensities with ease.

Earthquake intensities are also expressed in a logarithmic scale called the *Richter scale*. When earthquake force goes up one point in the Richter scale, actual energy released goes up 10 times.

Story points are similar to these logarithmic scales. The story point can accommodate a wide range of practical software sizes and present them in single digit numbers.

Classic Metrics: Unpopular Science

Classic software metrics have become an unpopular science, at least in parts, because of the evolution of software engineering. A few rigorous metrics have rendered truth less accessible, and such ill-designed metrics turned out to be masks; instead of revealing, they hid truth. Ambiguous metrics have made things worse being subject to multiple interpretations. The majority of classic metrics are not direct.

Effort variance, for example, is seldom based on true effort spent but is based on permissible numbers. An engineer may have spent 14 hours in a day but is asked to enter only 8 hours because that is the billable amount and the client is not paying for overtime. In this case, metrics breed hypocrisy, the very evil it seeks to fight. Users need to "understand" the hidden context and guess the hidden meaning.

Schedule variance in a business governed by a service-level agreement has no meaning but is still mentioned in the metric plan. Numbers are entered to fill reports; no one from the project takes such metric report seriously.

Productivity metrics and corporate goals on productivity are a source of perennial conflict. A common goal is set for all categories of tasks, unmindful of the differences in the underlying engineering principles and process capabilities. In such a case, no one believes productivity data.

Programmers have strong reasons for not believing the very definition of certain metrics. Managers have reasons for not trusting metric data; they fear fudging. Life with classic metrics goes on with deeply running distrust. Pressure for CMMI certification has pushed many organizations into this self-defeating situation. Because auditors insist on well-behaved data as evidence of good process control, people remove "outliers"— and truth. Errors in metric data are eliminated, destroying opportunities for learning.

> *Pressure is on metrics to present a perfect picture of life, in a totally unscientific manner.*

> *Metrics, when misused, establish illusions of perfection and scientific superstitions.*

There have been exceptions. There are genuine metric users, but they are far too few to turn the tide of opinion on metrics.

I believe in metrics, not politics.

Narayana Murthy
Chairman, Infosys

BOX 9.2   MEDICAL ANALOGY: MRI SCAN AND PULSE READING

A doctor asks for an MRI scan of the head to treat a headache. He wants to eliminate possible cause: clots in the brain. The MRI scan is a costly and elaborate procedure. The doctor is going through a causal analysis. The scan is normal, and the doctor pursues further causes of the headache, with the cost paid by the patient through trial and error. Finally, the patient could not be cured of his headache.

The patient decides to take naturopathy treatment. The skilled doctor reads his pulse, understands the problem, and suggests physiotherapy exercises and yoga for a month. The patient is cured.

Agile metrics are similar to reading the pulse. Data are cheap, simple, and quick; the solution is self-healing. The pulse reading reveals more to a trained doctor than the MRI scan.

Two Sides of Classic Metrics

Troubles notwithstanding, classic metrics have made a point. People have learned to use numbers, at least when they want to, first by imitating best practices and later by acquired capability. Classic metrics have always had great potential (untapped though). Those organizations that followed Personal Software Process (PSP) metrics did benefit: they reduced defects. Humphrey's vision of discipline from data worked on the one side; organizations showed Return on Investment (ROI). On the other side, PSP metrics were not popular and were resisted by teams. There were two sides to PSP metrics, the good and the bad. People wanted simpler solutions and pushed aside PSP metrics.

Automated metrics survived, which depend on tools, and people were asked just to use data not to collect data. That by itself made a huge difference. The drawback is that nobody is motivated to use data collected by tools. A seeming gain became a loss. Such data did not connect well with people except in the form of reports and business intelligence; the intended recipients were senior managers and customers. Designers and testers did not care.

Simple core metrics were good and they worked. Fancy special metrics did not. Metrics planned and metrics used were worlds apart. At least metric helped in goal setting. Process control loop using metrics were too circuitous and operated with time lags that by the time reports came projects were finished. Even the core metrics did not reach teams in time. The feedback loop disappeared on the way. Care was taken to install a metric, but no one thought about human communication.

Without communication links with people, metrics fail.

Metrics for Agile: Humanization

Agile projects have put people before processes and humanized metrics. Agile metrics instantly reach people through displayed charts and daily meetings. Metrics themselves are kept light, user-friendly, and direct. Code complexity is measured as high, medium, or low instead of using the sophisticated McCabe complexity number. Software sizes come in story points; one does not have to count function points using time-consuming rules by the International Function Point Users Group. More than that, metrics are readily used. Graphical elements have become signifiers instead of numerical indicators, where possible. The burn-down chart has replaced several metrics at once. Agile metrics influence day to day decision making.

The Price of Humanization

Agile metric data are mostly in the ordinal scale (like in "high–medium–low" judgments). Ordinal data can be collected quickly without hassle. Ordinal data also have a degree of calculated approximation. The contrast is between precise but less used classic metric data versus approximate but humanized agile data. Working with a lower scale suffers information loss, a price we pay for humanization. However, that loss does not make agile metrics less scientific than classic metrics, if users are aware of the degree of approximation.

> The aim of science is not to open the door to infinite wisdom,
> but to set a limit to infinite error.

Bertolt Brecht
Life of Galileo

> Sometimes a clearly defined error is the only way to discover
> the truth.

Benjamin Wiker
The Mystery of the Periodic Table

Common Agile Metrics

Velocity

Agile velocity is story points delivered per sprint. This is the rate at which the team delivers tested features. To obtain the best out of this metric, sprints (or iterations) must be of consistent length.

Story Point

This is an agile metric of software size, an agile counter part of the classic function point. Story point is a numerical expression of textual metrics of size: very small, small, medium, large, very large, and so on. Because human judgment is in a non-linear scale, we prefer a nonlinear order of numbers: 1, 2, 3, 5, 8, 13, 21, 34, 55, 89, 144, 233,..., the Fibonacci series. In the Fibonacci scale, the following conversion table can be used to express software size.

Software Size	Story Point
Very low	1
Low	2
Medium	3
Large	5
Very large	8
Extremely large	13
Next level	21

Software requirements are broken into small features called *user stories*. The story point scale is used to judge the size of "user stories."

User stories are analogous to the classical use cases.
Story Points are analogous to Use Case Points.

Technical Debt

Number of defects discovered per iteration.

Tests

The number of tests that have been developed, executed, and passed to validate a story.

Level of Automation

The percentage of tests automated.

Earned Business Value (EBV)

Business value attached to stories delivered. According to Dave Nicolette, "EBV may be measured in terms of hard financial value based on the anticipated return on investment (ROI) prorated to each feature or user story."

Burn-Down Chart

Burn down represents the remaining work of the project versus the remaining human resources. This information can be presented every week until a complete release. Burn-down chart is a famous agile visual used to track progress. Figure 9.1 shows an example of a burn-down chart.

A common practice is to plot a burn-down chart for every team and for each iteration. This provides the necessary biofeedback to teams to control backlog and to attain iteration goals in time.

Burn-Up Chart

Burn up represents work finished. Figure 9.2 illustrates a burn-up chart (BUC).

The *y*-axis is a cumulative plot of stories developed iteration by iteration. The ideal performance line is plotted along the actual performance line to provide guidance and to measure performance gaps. The BUC can be used when several stories are developed concurrently. This is a chart of stories built and work done; this records achievement and not merely activities.

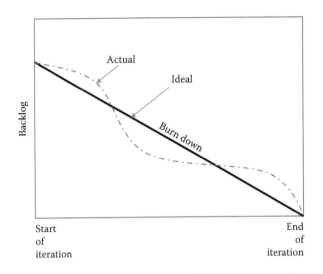

Figure 9.1   Burn-down chart.

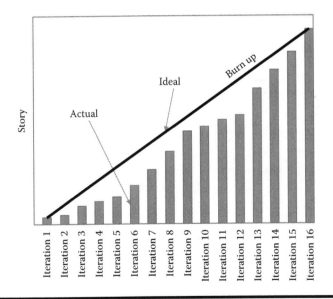

Figure 9.2   Burn-up chart.

Burn Up with Scope Line

A BUC with scope line marked above as shown in Figure 9.3 has an advantage.

Changing scope can be portrayed in this form and is not so easy in the other two charts.

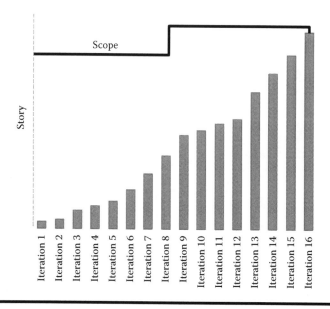

Figure 9.3 Burn-up chart with scope line.

BOX 9.3 A REBIRTH

Many classic metrics are reborn into the agile world with new names. Productivity used to be measured as KLOC/man month. This metric is called *velocity* and measured as story points delivered per iteration. The basic concept remains the same, but the metrics used in the calculation are now less precise but more convenient. The expectations have shifted. Velocity metric is not used as a target, and that seems to have made all the difference; the metric has received social acceptance.

When a measure becomes a target it ceases to be a good measure.

Goodharts Law

Good old defects are now called *technical debt*. For long, programmers knew that there should be no stigma attached to software defects. The new definition upholds a dignity and the human spirit.

Adding More Agile Metrics

The present metric system is a victim of poor implementation. It is overdesigned but underutilized. Agile metrics are simple and easy to implement; they have to be simple to honor the very spirit of agile methodology.

Hartmann and Dymond [1] list 10 attributes of a typical agile metric system as follows:

1. It affirms and reinforces lean and agile principles.
2. It follows trends, not numbers. Measure "one level up" to ensure you measure aggregated information, not suboptimized parts of a whole.
3. It belongs to a small set of metrics and diagnostics. A "just enough" metrics approach is recommended: too much information can obscure important trends.
4. It measures outcome, not output.
5. It is easy to collect.
6. It reveals rather than conceals.
7. It provides fuel for meaningful conversation.
8. It provides feedback on a frequent and regular basis.
9. It measures value.
10. It encourages "good enough" quality.

It may be noted that the above attributes can be applied to metrics in conventional life cycles too.

Hartmann and Dymond conclude that the key agile metric should be business value, and they note,

> Agile methods encourage businesses to be accountable for the investment in software development efforts. In the same spirit, the key metrics we use should allow us to measure this accountability. Metrics should help validate businesses that make smart software investments and teams that deliver business value quickly.

ROI begins with the first release of feature. Value must be measured. Delivering value early is the hallmark of agile projects.

However, in extreme programming, project teams have added metrics that are not so simple. Teams use the following metrics where refactoring takes place at the end of each iteration source code:

1. Coupling
2. Cyclomatic complexity

An example of value generated by these metrics is available in a case study by Martin Iliev [2]. Coupling metrics lead to "good encapsulation, high level of abstraction, good opportunity for reuse, easy extensibility, low development costs and low maintenance costs." Further, cyclomatic complexity metrics lead to "low maintenance costs, collective code ownership, easy to test and produce good code coverage results." Martin Iliev has established a firm business case for these metrics.

> *From the above example, it may be seen that metrics are agile because of the way they are used and the value they create and not because of their internal characteristics.*

In yet another case, Frank Maurer and Sebastien Martel [3] study productivity in extreme programming in OO projects using the following four metrics:

1. LOC/effort
2. Methods/effort
3. Classes/effort
4. (Bugs + features)/effort

They present evidence for improvement in productivity after introducing XP using the four metric data, a fairly obvious use of agile metrics to find ROI of process improvement. It may be noted that they have considered the metric productivity instead of velocity in this case study.

Case Study: Earned Value Management in the Agile World

BUCs in agile projects remind us of the earned value graph (EVG) in conventional projects. BUC and EVG look alike. The similarity runs deeper. Earned value management is widely accepted as a best practice in project management and is covered well in *Project Management Body of Knowledge*. Managing milestones makes a manager agile in sharp contrast with one who chooses to manage at the task level. There are typically about eight milestones in a project, and all the project manager had to do is to monitor earned value, planned value, and cost at every milestone and connect the dots and plot EVG. As milestones pass by, the project manager is able to predict future performance by seeing trends. A BUC does exactly that. We use sprints instead of milestones. Value is measured in terms of finished and tested stories.

The implementation of EVM in agile projects is explained by John Rusk [4], who observes, "Agile and EVM are a natural fit for each other."

Anthony Cabri and Mike Griffith [5] explore EVM usage in agile projects, create examples of BUCs, and tackle the issue of changing scope with EVM.

Tamar et al. [6] consider EVM in Scrum and find that "the implementation of the Agile EVM process has no noticeable impact on a Scrum team's velocity. Also, the value of the data was confirmed by the team who had access to the metrics, as well as the Scrum Master and management stakeholders for the project. Thus, we are encouraged that the Agile EVM metrics do indeed add value to Scrum projects."

In a study of EVM in waterfall and agile projects, Sam Ghosh [7] concludes that "the concept of EVM is applicable in agile software project."

BOX 9.4   RANK THE MESSAGE

Sometimes the message is not in absolute values but in the relative order. To transform nonnormal data into normal data, we consider ranks, for example. When there is deterioration in data, ranks still hold true. A dentist gives times to five of his clients: 10, 10:30, 11, 11:30, and 12:00. Patients come and wait, but the doctor arrives 30 minutes late. The first patient complains she has lost 30 minutes. The desk operator is cool and says, "You are the first patient the doctor will see. You are first in the queue. Ignore the actual time promised to you. But we will maintain the order." The patient reflects upon this response. The dental clinic is committed not to the time schedules but to the order. That is how the clinic sees it. Their message seems to be in the order, or the rank. The exact values of numbers are lost; the ranks are remembered and retained. It requires a greater disciple to remember exact values.

Review Questions

1. How many metrics would you use in a test dashboard in a testing project?
2. What are the metrics that can be used in unit testing?
3. What are the metrics that can be used in smoke tests?
4. Suggest simple ways of assessing reliability of software before release.
5. What metric data will be used while making a decision about stopping testing?

Exercise

1. Develop a minimum set of metrics you would maintain for a testing agency.

References

1. D. Hartmann and R. Dymond, *Appropriate Agile Measurement: Using Metrics and Diagnostics to Deliver Business Value*, IEEE, Agile Conference, 2006.

2. M. Iliev, I. Krasteva and S. Ilieva, *A Case Study on the Adoption of Measurable Agile Software Development Process*, 2009.
3. F. Maurer and S. Martel, *On the Productivity of Agile Software Practices: An Industrial Case Study*, 2002.
4. J. Rusk, *Earned Value for Agile Development*, The DACS, 2009.
5. A. Cabri and M. Griffith, *Earned Value and Agile Reporting*, 2006.
6. T. Sulaiman, B. Barton and T. Blackburn, Agile EVM—Earned Value Management in Scrum Projects, *AGILE '06: Proceedings of the Conference on AGILE 2006*, IEEE Computer Society, July 2006.
7. S. Ghosh, Systemic comparison of the application of evm in traditional and agile software project, *PM World Today* 14(2), 2012.

Suggested Reading

Aczel, A. D., *Complete Business Statistics*, McGraw-Hill/Irwin, London, 2005.

LAWS OF PROBABILITY

In Section III, we shall see some time-tested laws of statistics. However, the foundation lies in empirical patterns of evidential data because all these laws are inspired by patterns in evidential data.

Chapter 10 on histogram establishes the rules of thumb of pattern extraction from data. Chapter 11 presents the binomial process, the binomial and related laws. Historically, a good deal of statistical concepts evolved around these laws, both to conceive them and to apply them. Chapter 12 presents the exponential law, the related Poisson process. Chapter 13 presents the bell curve, which has become a social law of universal status and great potential. Statistical thinking is governed by the bell curve. The uniform and triangular bounded distributions, presented in Chapters 14 and 15, represent another set of laws that help business decision making. The well-known Pareto law, along with some less known applications in open source software development, is presented in Chapter 16.

These laws have resulted in several other laws and paradigms. They have cast permanent influence on several fields of science and engineering. They are also behind some of the most powerful knowledge systems and prediction models used in management. These laws are basic to the understanding of software engineering and management. They have inspired scientists, economists, and managers ever since their discovery. With time, the application and usage of these laws seem to increase exponentially.

Chapter 10

Pattern Extraction Using Histogram

The histogram is easily the most commonly used statistical method. It is used to detect frequency patterns in data.

A histogram is a way to count the number of data points in data intervals. First, the data range that extends from the minimum to maximum value is divided into a certain number of equal intervals. Then we count the number of data points in each interval and tabulate the counts in a table called *tally*. The frequency table is converted into a bar graph known as a *histogram*. An example of data creating a tally, constructing a histogram, and deriving a frequency diagram as well as an ogive from the histogram is illustrated in Appendix 10.1.

A histogram of requirements volatility is shown in Figure 10.1. This histogram is similar to the one obtained by Kulk and Verhoef [1], who studied requirements volatility in 84 IT projects comprising 16,500 function points. Requirement changes do vary beyond the traditional limit of 10%.

What draws our attention first in the histogram is its peak. Stable processes have strong peaks. The peak represents the mode of the process. The core process is denoted by the body of the histogram. Almost the entire process results are seen in the core. Outside the core, we can see outliers. Unlike in the box plot, outliers are distinguished by contrasts in the pattern and not by any rules. This histogram is symmetrical. Many metrics exhibit nearly symmetrical shapes; effort variance and schedule variance are well-known examples.

Not every histogram is symmetrical. For example, the histogram of complexity of object oriented structures, measured as weighted methods per class, is shown in Figure 10.2, modeled after the finding of Rosenberg [2].

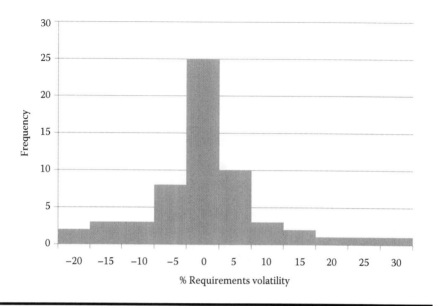

Figure 10.1   Symmetrical histogram of requirements volatility.

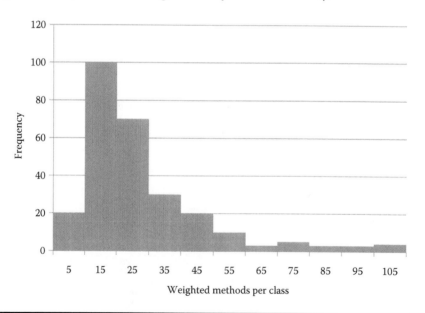

Figure 10.2   Skewed histogram of object complexity.

This histogram is skewed. There are few classes with large weighted methods per class. The larger the number of methods in a class, the greater the potential effect on children; children inherit all the methods defined in the parent class. Classes with many methods are likely to limit the possibility of reuse. They are also difficult to understand and test. Those few complex objects require review and a relook. A lot more metrics show skewed shapes; time to repair, defect density, and productivity are known for their skew.

BOX 10.1   HISTOGRAM IN CAMERAS

Histogram is a digital signature of reality. It is used in modern digital cameras to present a pictorial view of light intensity in the field of view. The light profile of objects seen through the lens is scanned, digitized, and converted into a histogram. The photographer can derive clues from this histogram to the settings required to get a good picture.

Understanding image histograms is probably the single most important concept to become familiar with when working with pictures from a digital camera. A histogram can tell you whether or not your image has been properly exposed, whether the lighting is harsh or flat, and what adjustments will work best. It will not only improve your skills on the computer, but as a photographer as well. (http://www.cambridgeincolour.com/tutorials/histograms1.htm)

Before the histogram, photography enthusiasts had to go through a lot more effort to get good exposures.

Image editors typically have provisions to create a histogram of the image being edited. The histogram plots the number of pixels in the image (vertical axis) with a particular brightness value (horizontal axis). Algorithms in the digital editor allow the user to visually adjust the brightness value of each pixel and to dynamically display the results as adjustments are made. Improvements in picture brightness and contrast can thus be obtained. (http://en.wikipedia.org/wiki/Image_editing)

Choosing the Number of Intervals

Square Root Formula

The elements of a histogram, namely, the peak, the body, and the outliers, can change if we change the number of interval N. Normally, we choose the number of intervals to be the square root of the number of data points in the sample n.

$$N = n^{0.5} \tag{10.1}$$

This square root formula is taken as a default value in histogram analysis.

Alternate Approaches

There are three other conventions as well in selecting the number of intervals.

1. The Sturges rule,

$$N = \log_2 n + 1 \tag{10.2}$$

2. The Freedman–Diaconis rule,

$$N = \frac{\text{Range}}{2IQR(n - 1/3)} \tag{10.3}$$

3. The Scott rule, which suggests fewer intervals,

$$N = \frac{\text{Range}}{3.5\sigma\sqrt[3]{n}} \tag{10.4}$$

Exploratory Iterations

We can see how the histogram varies when the number of bin intervals—and as a direct consequence the bin sizes—vary according to the four rules previously mentioned. Our trials need not be limited to these four rules. We can try our own choice of bin size and extract patterns to suit our inquiry. Bin size reduction increases the resolution of histogram graph. For example, in the first iteration, with just nine bins, effort variance data yield a histogram shown in Figure 10.3. The histogram shows stability and a single peak. We can explore further by improving the resolution of the histogram and choose 20 bins; we get a histogram shown in Figure 10.4.

This histogram has three modes, or three clusters. This is merely an estimate. Extracting different histogram estimates with the same data set is known as *nonparametric density function estimation*.

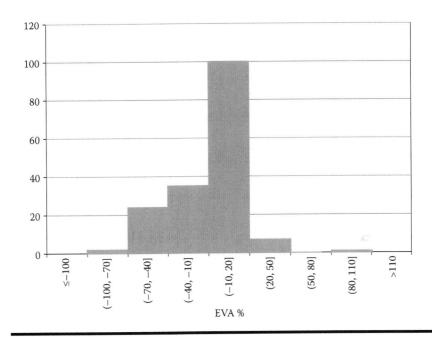

Figure 10.3   Histogram of effort variance (EVA%) with nine bins.

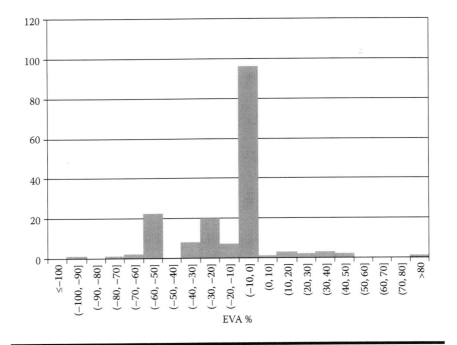

Figure 10.4   Histogram of effort variance (EVA%) with 20 bins.

BOX 10.2   HISTORY OF HISTOGRAMS

The word "histogram" is of Greek origin, as it is a composite of the words "istos" (= "mast") and "gram-ma" (= "something written"). Hence, it should be interpreted as a form of writing consisting of "masts," i.e., long shapes vertically standing, or something similar. The term "histogram" was coined by the famous statistician Karl Pearson to refer to a common form of graphical representation. Histograms were used long before they received their name, but their birth date is unclear. It is clear that histograms were first conceived as a visual aid to statistical approximations. Bar charts most likely predate histograms and this helps us put a lower bound on the timing of their first appearance.

Yannis Ioannidis
Department of Informatics and Telecommunications, University of Athens

Process Signature

> Our humanity rests upon a series of learned behaviors, woven together into patterns that are infinitely fragile and never directly inherited.
>
> **Margaret Mead**

Data of process performance can be converted into histogram signatures. These process signatures represent process characteristics. They are used to manage processes.

1. *Process stability*: Histogram can be applied to test process stability. Stable processes produce histograms with a single peak.
2. *Mode*: Histogram reveals process mode.
3. *Multiple peaks*: If data come from a mixture of several processes, we will see multiple peaks in the histogram.
4. *Cluster analysis*: If data have natural clusters, histograms show them.
5. *Outliers*: Histogram can easily show outliers.
6. *Natural boundary*: Histogram reveals natural process boundaries that can be used in goal setting.

The histogram in Figure 10.5 is an example of process signature. It captures the way time to repair is managed in projects and presents a broad summary of historic performance.

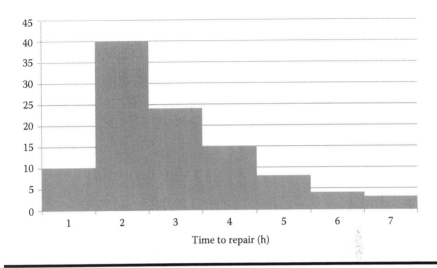

Figure 10.5   Process signature histogram of time to repair.

This is a typical experience in fixing high-priority bugs without any SLA con-straint. The team treats high-priority bugs with utmost earnestness and tries to ship the fix at the earliest. The histogram is skewed and has a thick tail on the right side. There is a sure probability that the repair time would be high.

Beaumont [3] shows a more disciplined histogram for time to repair. The dis-persion is far less. Barkman et al. [4] present histograms for 16 selected metrics in open source projects. (Data have been collected from 150 distinct projects with over 70,000 classes and over 11 million lines of code.) This is really a gallery of histogram signatures; entries range from well-behaved symmetrical histograms to extremely skewed ones. The signature structures typically represent the metrics. These patterns are more or less the same across the entire IT industry.

Uniqueness of Histogram Signature

Histograms are true signatures.

> *Process histograms reflect people.*
> *Product histograms reflect design.*

People leave their signatures in their deliveries. The uniqueness of histograms can be used to advantage. Software Engineering Institute (SEI) has presented a series of histograms that change with the maturity of the organization from level 2 to level 5 in their several communications. Process histogram is a signature of an organization's maturity. Well-constructed histograms with the right metrics can be used as more precise signatures that can be used to compare and predict performance.

A brilliant case in point is the video signature of Liu et al. [5]:

> The explosive growth of information technology and digital content industry stimulates various video applications over the Internet. Duplicate detection and measurement is essential to identify the excessive content duplication. There are approximately two or three duplicate videos among the ten results on the first web page. Finding visually similar content is the central theme in the area of content-based image retrieval; histogram distributions of similar videos are with much likeness, while the dissimilar ones are completely different. The video histogram is used to represent the distributions of videos' feature vectors in the feature space. This approach is both efficient and effective for web video duplicate detection.

BOX 10.3   DETECTING BRAIN TUMOR WITH HISTOGRAM

Brain cancer can be counted among the most deadly and intractable diseases. Tumors may be embedded in regions of the brain forming more tumors too small to detect using conventional imaging techniques. Malignant tumors are typically called *brain cancer*. These tumors can spread outside of the brain. Brain tumor detection is a serious issue in medical science. Imaging plays a central role in the diagnosis and treatment planning of a brain tumor.

The image of the brain is acquired through MRI technique. If the histograms of the images corresponding to the two halves of the brain are plotted, a symmetry between the two histograms should be observed due to the symmetrical nature of the brain along its central axis. On the other hand, if any asymmetry is observed, the presence of the tumor is detected. After detection of the presence of the tumor, thresholding can be done for segmentation of the image. The differences of the two histograms are plotted and the peak of the difference is chosen as the threshold point. Using this threshold point, the whole image is converted into a binary image providing the boundary of the tumor. The binary image is now cropped along the contour of the tumor to calculate the physical dimension of the tumor. The whole of the work has been implemented using MATLAB® 2010. (Kowar and Yadav [6])

Histogram Shapes

Histograms are empirical distributions (or density functions). They can be smoothed by nonparametric methods, as is performed in machine intelligence algorithms. Alternatively, they can be fitted to mathematical models.

The shape of a histogram can help in deciding suitable mathematical equations. Skewed histograms suggest lognormal, exponential, or Pareto distributions. The length of a histogram tail contains the final clues. Symmetrical histograms suggest normal distribution. Left tails suggest Gumbel minimum distribution. Abrupt right tails suggest Gumbel maximum distribution. (The previously mentioned mathematical distributions are described in Section II of this book.) These descriptions referred to the 16 histograms presented by Barkman [7], which can be visually mapped to well-known probability distributions. A visual selection of the best suited equation clue is a valuable low-cost alternative to complex techniques for model building.

Mixture

If a histogram has a second peak, it is known as *bimodal*. The second peak (or cluster) may come from a mixture of data from two processes. For example, a productivity histogram may exhibit two peaks. Each peak may correspond to one programming language containing a mixture of data. Alternatively, the better peak in the productivity histogram may come from a different team performing with higher skill levels, and that is the case with the histogram shown in Figure 10.6. The way histograms reveal mixtures is very helpful.

Process Capability Histogram

> Character is expressed through our behavior patterns, or natural responses to things.
>
> **Joyce Meyer**

Although the process presented results in histograms, it is customary to mark the upper specification limit (USL) and the lower specification limit (LSL) on the histogram. With USL and LSL marks, it is now called a *process capability histogram*. It enables us to check if the process peak is on target and if the process variation is within the limits. These two are the criteria for a capable process. Process capability indices can be calculated along with process risk.

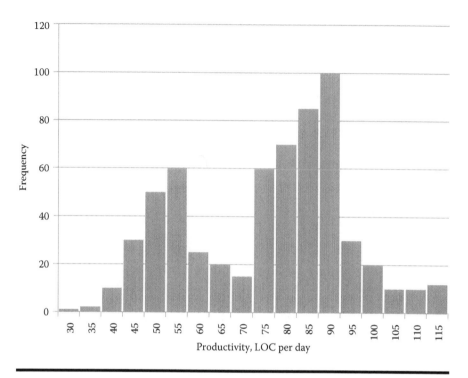

Figure 10.6 Bimodal histogram of productivity.

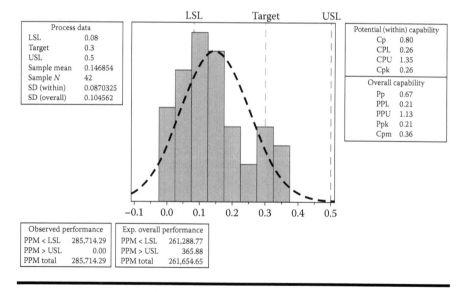

Figure 10.7 Process capability of defects/test case.

Figure 10.7 is a histogram of defects found per test case, a key metric in software testing. The LSL has been set at 0.08, and the USL has been specified at 0.5. Perhaps specifying a USL is unnecessary, but it is meant to cause an alert on product quality in this case.

The lower limit is less controversial; the team expects a minimum return from test cases. The target 0.3 defects found per test case is also marked for reference. Visually, the histogram reveals capability-related issues: the process peak is not on target, and a good deal of results (roughly 28%) fall below the lower specification limit. The process has a lot to improve. The capability index is 0.26, much lower than the desirable value of at least 0.8.

The test case metric "defects found per test case" is a double-edged sword. If more defects are found, it could be either due to the poor quality of product on test or highly effective test cases. If few defects are found, it could mean that the product is good or the test cases are not effective. One has to appreciate both the possibilities and use extraneous evidence to judge the histogram.

Histogram as a Judge

There is a very significant use of the histogram as a judge. First, it is a visual judge of the normality of data. There are debates about the normality of metric data, and people do a normality test. A visual judgment of the histogram of the data can be a first-order judgment of normality. If data are not normal, people do not do esoteric statistical tests on the data.

Process data in software development is often nonnormal.

In these cases, the histogram is used to visualize data and to make a decision about statistical tests. For example, time to repair data are avowedly nonnormal; all known histograms testify to this. The problem is now escalated: one should use nonparametric tests, or one should transform data appropriately and do statistical tests. The author prefers the first. Let us keep data in its purest form.

There is an area where we are sure data will have to be normal: prediction errors. In any prediction model, errors are symmetric around the mean, and the mean error value is zero. A histogram is of the errors usually plotted, to see if it is symmetrical around zero, to validate the prediction model.

Good regression models leave behind errors, or residuals, which are normally distributed and can be tested with histograms.

If the histogram is skewed, the model is not accepted and needs to be improved.

> **BOX 10.4   WEB CONTENT EXTRACTION**
>
> The content of web pages is extracted by using the HTML document's Text-To-Tag Ratio histograms. Web content extraction is seen as a histogram clustering task. Histogram clustering is a widely researched topic that is especially popular with image researchers. This is especially true among researchers who wish to use the histogram footprints of images as a means for classification, segmentation, etc. These clustering techniques are also enhanced with the use of histogram smoothing techniques. High recall and precision are achieved by this technique. (Weninger and Hsu [8])

From One Point to One Histogram

To judge a process or a product, a single observation is not enough. We need multiple observations and a histogram constructed out of the multiple observations. The minimum element in statistical structure seems to be a histogram. If we have to judge a process, we need to make enough observations and plot a process capability histogram. If we have to study product behavior, we need to collect enough data from a minimum number of modules and construct a histogram. Both the process and the product can be observed only through histograms.

Case Study: Goal Entitlement

Setting stretch goals is a tricky challenge. Such goals have to be realistic too. The goal setting process must be transparent as well. This case study presents the use of histograms in setting stretch goals.

Productivity analysis using histogram reveals the presence of two clusters. The stronger peak seems to be of higher productivity. The smaller cluster is closer to the present goal of 50 lines of code (LOC) per day.

There is every reason to increase the goal to the best practice cluster peak, marked B, following a natural path of improvement shown in Figure 10.8.

> *Entitlement is the best you can possibly operate without redesigning your process. It is the difference between the current level of performance and the best documented.*

This is the core concept in "goal entitlement." A conservative stretch goal will be to set any intermediate point C on the path of improvement such that 70% of the ideal improvement is targeted. Thus, C becomes a new goal. With the histogram in the background, the entire analysis and planning is data based and realistic and makes it easy for people to accept the stretch goal without any reservation.

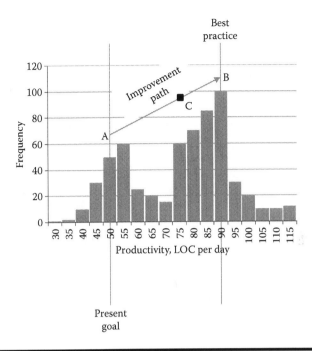

Figure 10.8   Histogram representing an improvement path.

Goal entitlement is contested by some, as in *The Entitlement Trap* by Dennis J. Monroe [9]. He argues, "Goal setting by entitlement does not quantify opportunity for breakthrough improvement."

BOX 10.5   A STUDY OF REUSE

The benefits of software reuse range from decreased development time and increased product quality to improved reliability and decreased maintenance costs. The amount of reuse is measured by several metrics. Curry et al. [10] focused on three metrics:

1. *Reuse level (RL)*: It measures the ratio between different lower level items reused verbatim inside a higher level item versus the total number of lower level items used.
2. *Reuse frequency (RF)*: It measures the number of references to reused items rather than counting items only once, as was performed for the reuse level. This metric measures the percentage of references to lower level items reused verbatim inside higher level items versus the total number of reference.

3. *Reuse density (RD)*: It measures how much reuse is in a product with respect to the size of the product. It is the ratio of total number of lower level items reused inside a higher level item normalized to the size of the higher level item.

The question is if the three metrics are redundant. To solve the problem, the authors have used histograms of correlations. The analysis shows interesting findings with statistical confidence made available by histograms. Instead of using an average level of correlation, the authors have preferred histogram expressions of correlations. They conclude,

> It is evident that from a statistical point of view in the considered C projects, RL and RF measure very similar properties of the code, while RD presents an independent perspective on the amount-of-reuse.

Just two metrics are enough to manage reuse.

Appendix 10.1: Creating a Histogram

There are a few standard steps used in histogram creation. We are using the Excel–Data–Data Analysis–Histogram option.

Data are made available as a column. Column A in Figure A10.1 contains coupling data between objects. Many objects seem to be self-contained but quite a few are coupled (Data A10.1).

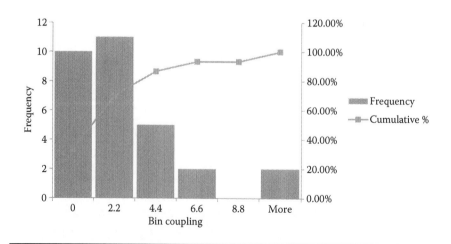

Figure A10.1   Histogram and ogive (cumulative %).

Data A10.1   Coupling Data

1	0
10	4
2	0
3	2
2	0
3	5
2	0
1	2
1	0
2	6
0	0
0	11
1	0
0	4
3	1

Table A10.1   Tally

Bin	Frequency	Cumulative %
0	10	33.33
2.2	11	70.00
4.4	5	86.67
6.6	2	93.33
8.8	0	93.33
More	2	100.00

We run the tool Histogram. Define the input range and select the option for chart output and cumulative percentage. Take the report from Excel by specifying "output range." Table A10.1 and Figure A10.1 show the histogram of coupling and the cumulative curve, known as *ogive*. Select tally columns Bin and Frequency and plot an XY scatter diagram to obtain Figure A10.2. This figure is a smoothened profile of the histogram.

Interpretation

The coupling histogram is skewed to the right. The tendency of developers seems to produce less complex classes. That is a very good sign. There seems to be a few high complex outlier classes. This appears as a small independent bar on the right.

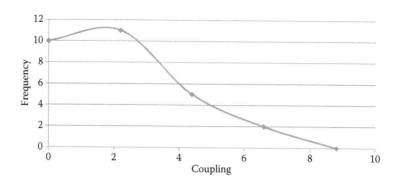

Figure A10.2 Frequency diagram.

 ## Review Questions

1. What is the commonly used rule for selecting number of bins in a histogram?
2. Mention three purposes for extracting histograms from data.
3. What is meant by a bimodal histogram?
4. What are the elements of a histogram signature?
5. How can you judge process capability from a process data histogram?

Exercises

1. Construct a histogram using the following customer satisfaction data and extract the signature of customer satisfaction. Interpret the signature.

    4, 5, 3; 5, 4, 5, 5, 4, 3, 4, 5, 3, 4, 5, 3, 2, 3, 2, 4, 1, 4,
    1, 5, 4, 3, 4, 3, 2, 3, 2, 4, 3, 4, 3, 5, 4, 3, 5, 4, 4, 4, 4

2. Use MS Excel's data analysis tool Histogram to construct the histogram. Instead of allowing default bin selection, specify your own bins.
3. If the corporate goal is to get a customer satisfaction score of at least 3, what is the risk seen in the previously mentioned histogram signature?
4. Test case rework effort data in person-hours are given as follows:

    16, 16, 2, 5, 7, 8.5, 8, 9, 10, 11, 5

    Draw a histogram with this limited data and try to draw inferences about the test case development process.
5. Effort variance data in a software enhance project is shown as follows:

    −10, 4, 5, −3, 8, −2, 0, 9, 5, −2, 5, 3, 5, 12

    Draw an ogive of the given data.

References

1. G. P. Kulk and C. Verhoef, Quantifying requirements volatility effects, *Science of Computer Programming*, 72, 136–175, 2008.
2. L. Rosenberg, Applying and interpreting object oriented metric, *Software Technology Conference*, Utah, April 1998.
3. L. R. Beaumont, *Metrics—A Practical Example*, AT&T Bell Laboratories. Available at http://www.humiliationstudies.org/documents/Beaumontmetricsid.pdf.
4. H. Barkman, R. Lincke and W. Lowe, *Quantitative Evaluation of Software Quality Metrics in Open-Source Projects*, Software Technology Group, School of Mathematics and Systems Engineering, Sweden. Available at http://www.arisa.se/Files/BLL-09.
5. L. Liu, W. Lai, X.-S. Hua and S.-Q. Yang, *Video Histogram: A Novel Video Signature for Efficient Web Video Duplicate Detection*, Department of Computer Science and Technology, Tsinghua University, Microsoft Research, Asia, Springer Verlag Berlin, Heidelberg, 2007.
6. M. K. Kowar and S. Yadav, Brain tumor detection and segmentation using histogram thresholding, *International Journal of Engineering and Advanced Technology*, 1(4), 16–20, 2012.
7. W. E. Barkman, *In-Process Quality Control for Manufacturing*, CRC Press, New York, 1989.
8. T. Weninger and W. H. Hsu, *Web Content Extraction Through Histogram Clustring*, Kansas State University, Manhattan, Annie Conference, November 9–12, St. Louis, Missouri., 2008.
9. D. J. Monroe, *The Entitlement Trap*, Juran Institute, Inc., 2009.
10. W. Curry, G. Succi, M. Smith, E. Liu and R. Wong, *Empirical Analysis of the Correlation between Amount-of-Reuse Metrics in the C Programming Language*, ACM, New York, 135–140, 1999.

Chapter 11

The Law of Large Numbers

Let us move further away from frequency distribution and look at probability distributions. The frequency distribution that we have seen in Chapter 10 is an empirical pattern; what we are now going to see in this chapter and the rest of the book are mathematical expressions.

> The mathematical sciences particularly exhibit order, symmetry, and limitation; and these are the greatest forms of the beautiful.
>
> **Aristotle**

BOX 11.1   BIRTH OF PROBABILITY

Before the middle of the 17th century, the term *probable* meant approvable and was applied in that sense, univocally, to opinion and to action. A probable action or opinion was one such as sensible people would undertake or hold in the circumstances. However, the term probable could also apply to propositions for which there was good evidence, especially in legal contexts.

In the Renaissance times, betting was discussed in terms of odds such as "ten to one," and maritime insurance premiums were estimated based on intuitive risks. However, there was no theory on how to calculate such odds or premiums.

The mathematical methods of probability arose in the correspondence of Pierre de Fermat and Blaise Pascal (1654) on such questions as the fair division of the stake in an interrupted game of chance.

Fermat and Pascal helped lay the fundamental groundwork for the theory of probability. From this brief but productive collaboration on the problem of points, they are now regarded as joint founders of probability theory. Fermat is credited with carrying out the first ever rigorous probability calculation. In it, he was asked by a professional gambler why if he bet on rolling at least one six in four throws of a die he won in the long term, whereas betting on throwing at least one double-six in 24 throws of two dice resulted in his losing. Fermat subsequently proved why this was the case mathematically. (http://en.wikipedia.org/wiki/Problem_of_points)

Christiaan Huygens (1657) gave a comprehensive treatment of the subject.

Jacob Bernoulli's *Ars Conjectandi* (posthumous, 1713) put probability on a sound mathematical footing, showing how to calculate a wide range of complex probabilities.

Life Is a Random Variable

Results, in general, are random in nature; some could be in our favor and some not. Process results do not precisely remain favorable all the time, neither do they become unfavorable all the time. Results toggle between favor and disfavor, randomly.

> The measure of the probability of an event is the ratio of the number of cases favorable to that event, to the total number of cases.
>
> **René Descartes**

The discovery of probability goes back to the Renaissance times (see Box 11.1).

A process that toggles between favor and disfavor is called the *Bernoulli process*, named after the inventor. Mathematically, a Bernoulli process takes randomly only two values, 1 and 0. Repeated flipping a coin is a Bernoulli process; we get a head or tail, success or failure, "1 or 0." Every toss is a Bernoulli experiment. The Bernoulli random variable was invented by Jacob Bernoulli, a Swiss mathematician (see Box 11.2 for a short biography).

Results from trials converge to the "expected value" as the number increases. In an unbiased coin, the "expected value" of the probability of success (probability of appearance of heads) is 0.5. More number of trials are closer to the value of the probability of success. This is known as the *law of large numbers*. Using this law, we can predict a stable long-term behavior. It took Bernoulli more than 20 years to develop a sufficiently rigorous mathematical proof. He named this his *golden theorem*, but it became generally known as *Bernoulli's theorem*. This theorem was applied to predict how much one would expect to win playing various games of chance.

From sufficient data from real-life events, we can arrive at a probability of success (p) and trust that the future can be predicted based on this.

Prediction means estimation of two values: mean and variance (which denote central tendency and dispersion).

In this chapter, we consider four distributions to describe four different ways of describing the dispersion pattern.

1. Binomial distribution

   The probability of getting exactly k successes in n trials is given by the following binomial expression:

$$P(X = k) = C_k^n \, p^k (1 - p)^{n-k} \tag{11.1}$$

where n is the number of trials, p is the probability of success (same for each trial), and k is the number of successes observed in n trials, calculated as follows:

$$\text{Mean} = np \tag{11.2}$$

$$\text{Variance} = np(1 - p) \tag{11.3}$$

Equation 11.1 is a paradigm for a wide range of contexts. In service management, success is replaced by arrival, and the Bernoulli process is called *arrival-type process*. In software development processes, we prefer to use the term success. The coefficient C is a binomial coefficient, hence the name binomial distribution.

Software development processes may consist of two components:

a. An inherent Bernoulli component that complies with the law of large numbers
b. Influences from spurious noise factors

Bernoulli distribution is used in statistical process control. The spurious noise factors must be identified, analyzed, and eliminated. For example, in service-level agreement (SLA) compliance data, one may find both these components. If the process is restricted to the Bernoulli type, the process is said to be under statistical control. (Shewhart called this variation due to "common causes" and ascribed spurious influences to special causes.)

Equation 11.1 is used in the quality control of discrete events.

BOX 11.2 JACOB BERNOULLI (1654–1705)

Nature always tends to act in the simplest way.

Jacob Bernoulli

Jacob Bernoulli gave a mathematical footing to the theory of probability. The term Bernoulli process is named after him. A well-known name in the world of mathematics, the Bernoulli family has been known for their advancement of mathematics. Originally from the Netherlands, Nicolaus Bernoulli, Jacob's father, moved his spice business to Basel, Switzerland. Jacob graduated from the University of Basel with a master's degree in philosophy in 1671 and a master's degree in theology in 1676. When he was working toward his master's degrees, he would also study mathematics and astronomy. In 1681, Jacob Bernoulli met mathematician Hudde. Bernoulli continued to study mathematics and met world-renowned mathematicians such as Boyleand Hooke.

Jacob Bernoulli saw the power of calculus and is known as one of the fathers of calculus. He also wrote a book called *Ars Conjectandi*, published in 1713 (8 years after his death).

Bernoulli added upon Cardano's idea of the law of large numbers. He asserted that if a repeatable experiment had a theoretical probability p of turning out in a certain "favorable" way, then for any specified margin of error, the ratio of favorable to total outcomes of some (large) number of repeated trials of that experiment would be within that margin of error. By this principle, observational data can be used to estimate the probability of events in real-world situations. This is what is now known as the law of large numbers. Interestingly, when he wrote the book, he named this idea the "Golden theorem."

Bernoulli had received several awards and/or honors. One of the honors given to him was a lunar crater named after him. In Paris, there is a street named after the Bernoulli family. The street is called *Rue Bernoulli*.

Example 11.1: Binomial Distribution of SLA Compliance

QUESTION

From the previous year's deliveries, it has been estimated that the probability of meeting SLA in an enhancement project is 90%. Find out the probability of meeting SLA in at least 10 of 120 deliveries scheduled in the current year. Plot the related binomial distribution.

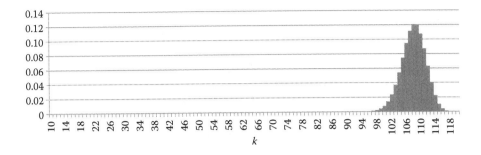

Figure 11.1   Binomial probability of SLA compliance (*p* = 0.9, *n* = 120).

ANSWER

You can solve Equation 11.1 by keeping p = 0.9, n = 120, and k = 10. This will give the probability of exactly 10 deliveries meeting the SLA. We are assessing the chance of getting exactly 10 successes.

Alternatively, use MS Excel function BINOM.DIST.

The answer is zero. The number is too small, 4.04705E-97.

Figure 11.1 shows the binomial probability distribution of SLA compliance. This may be taken as a process model. One can notice the upper and lower boundaries of the density function, approximately from 91 to 118 trials; one can also note the central tendency, which is exactly the mean.

BOX 11.3   TESTING RELIABILITY USING NEGATIVE BINOMIAL DISTRIBUTION

To test reliability, we randomly selected and run test cases covering usage. Executing a complete test library is costly, so we resort to sampling. We can choose inverse sampling and choose and execute test cases randomly until a preset number of defects are found (unacceptable defect level). If this level is reached, the software is rejected. Using regular sampling and under binomial distribution, we can do an acceptance test, but we might require to execute a significantly large number of test cases to arrive at an equivalent decision. Inverse sampling under NBD is more efficient in user acceptance testing.

With this information, we can construct the negative binomial distribution of defects. The salient overall point of the comparison is that, unless the software is nearly perfect, the negative binomial mode of sampling brings about large reductions in the average number of executions over the binomial mode of sampling for identical false rejection and false acceptance risks [1].

2. Negative binomial distribution

Negative binomial distribution (NBD) is defined by the probability of getting k successes until r failures occur, given by the following expression:

$$P(X = k) = C_k^{r+k-1} p^k (1 - p)^r \qquad (11.4)$$

where n is the number of trials, p is the probability of success (same for each trial), k is the number of successes observed in n trials, and r is the number of failures.

$$\text{Mean} = \frac{r(1 - p)}{p} \qquad (11.5)$$

$$\text{Variance} = \frac{r(1 - p)}{p^2} \qquad (11.6)$$

If k remains as an integer, the distribution is sometimes known as the *Pascal distribution*. Many engineering problems are elegantly handled with NBD.

In sampling, if the proportion of individuals possessing a certain characteristic is p and we sample until we see r such individuals, then the number of individuals sampled is a negative binomial random variable.

The NBD is one of the most useful probability distributions. It is used to construct models in many fields: biology, ecology, entomology, and information sciences [2].

Example 11.2: NBD of Right First-Time Delivery

QUESTION

In a network sensor manufacturing division, the right first-time rate is 0.6. The company wants to deliver 10 sensors to a mission critical application and prefers to ship after choosing from the right first-time lot. What is the probability of delivering 10 right sensors produced for the first time if the production batch size is 12? Plot the negative binomial probability distribution function associated with this problem. Calculate the mean and variance of the distribution.

ANSWER

It may be seen that data can be represented in Equation 11.4 with the following parameters:

$r = 10$ number of successes
$p = 0.6$ probability of success
$k = n - 10$ number of failures
$n =$ production batch size, 10, 11, ...

We can use the Excel function NEGBINOM.DIST to generate the NBD and plot the graph, as shown in Figure 11.2.

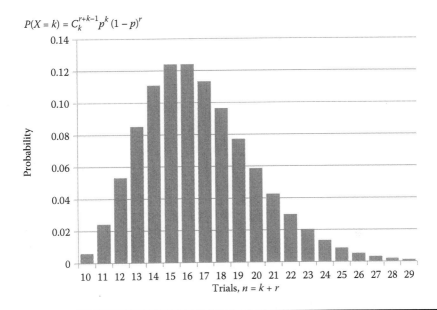

$$P(X = k) = C_k^{r+k-1} p^k (1 - p)^r$$

Figure 11.2 Negative binomial for right first time delivery ($r = 10$, success probability $p = 0.6$).

The Excel function appears as follows, with four arguments:

NEGBINOM.DIST(number_f, number_s, probability_s, cumulative)

where number_f is the number of failures (k in NBD Equation 11.4), number_s is the threshold number of successes (r in NBD Equation 11.4), probability_s is the probability of success (p in NBD Equation 11.4), and cumulative is a logical value that determines the form of the function. If cumulative is *true*, NEGBINOM.DIST returns the cumulative distribution function; if *false*, it returns the probability density function.

Finding the probability of delivering 10 right sensors produced for the first time from a batch of size 12 can be directly solved as follows:

Batch size $n = 12$
Number of success $s = 10$
Number of failure $k = 2$
Probability of success $p = 0.6$

The Excel function returns the answer 0.0532. Thus, there is only a small chance of finding 10 right sensors produced for the first time.

The mean and variance of the NBD can be directly computed by entering data in Equations 11.5 and 11.6. The answers are as follows:

Mean = 15
Variance = 37.5

Figure 11.2 shows that the sensor problem peaks at the mean.

3. Geometric distribution

The probability k of Bernoulli trials needed to obtain one success is given by the following expression:

$$P(X = k) = (1 - p)^{k-1} p \tag{11.7}$$

where p is the probability of success (same for each trial) and k is the number of successes observed in n trials.

$$\text{Mean} = \frac{1}{p} \tag{11.8}$$

$$\text{Variance} = \frac{1 - p}{p^2} \tag{11.9}$$

Example 11.3: Geometric Distribution

QUESTION

The right first-time design probability in a software development project is estimated at 0.7. Estimate the probability of needing four trials to find a defect-free feature design. Plot a graph between trials and geometric probability.

ANSWER

In this problem, $p = 0.7$ and $k = 4$.

Inserting these values in Equation 11.3, we obtain the geometric probability (0.0189).

Figure 11.3 shows the graph.

4. Hypergeometric distribution

The hypergeometric distribution is a discrete probability distribution that describes the probability of k successes in n draws without replacement from a finite population of size N containing exactly K successes. This is given by the following equations:

$$P(X = r) = \frac{C_r^K C_{n-r}^{N-K}}{C_n^N} \tag{11.10}$$

$$\text{Mean} = n \frac{K}{N} \tag{11.11}$$

$$\text{Variance} = n \left(\frac{N - K}{N} \right) \left(\frac{N - n}{N - 1} \right) \tag{11.12}$$

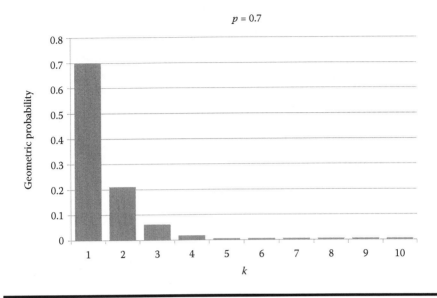

Figure 11.3 **Geometric probability distribution for defect free design.**

Example 11.4: Hypergeometric Probability

QUESTION

A release of 10 modules has just been built and the smoke test is over. Results show that there are four defective modules. If we draw samples of size 3 without replacement, find the probability that a sample contains two defective modules.

ANSWER

First, we assume that the proportion of defective modules follows the law of averages and holds good for every module. Given the fact that smoke tests do not find all defects, such an assumption has serious implications. However, to go ahead with solution formulation, we proceed with the following assumption:

You can solve Equation 11.4 by substituting $N = 10$, $K = 4$, $n = 3$, and $r = 2$

Alternatively, use Excel statistical function HYPGEOM.DIST to solve Equation 11.4. The data entry window must be filled as follows:

Sample_s   Number of successes in the sample   Enter 2

Note: Success in a statistical sense is finding a defective module. Testers also share this view.

Number_sample	Size of the sample	Enter 3
Population_s	Number of successes in the population	Enter 4
Number_pop	Population size	Enter 10
Cumulative	Logical value that determines the form of the function. If cumulative is *true*, then HYPGEOM.DIST returns the cumulative distribution function; if *false*, it returns the probability mass function.	Enter *false*

Excel returns the following answer: formula result = 0.3.

Thus, the probability that a sample of three modules contains two defective modules is 0.3.

Plots of Probability Distribution

To plot the PDF of hypergeometric probability, two scenarios are considered. The first is an inquiry into the chance of all items in the sample being defective. Figure 11.4a presents a plot between sample size and hypergeometric probability. The second is a study of one item in a sample being defective. Figure 11.4b presents the plot between sample size and hypergeometric probability.

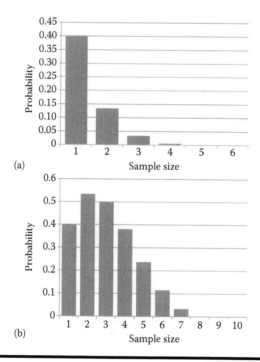

(a)

(b)

Figure 11.4 (a) Hypergeometric probability of all items in a defective sample (N = 10, K = 4). (b) Hypergeometric probability of one sample in a defective sample (N = 10, K = 4).

BOX 11.4   REVEREND THOMAS BAYES (1702–1761)

In 1719, Bayes matriculated at the University of Edinburgh where he studied logic and theology. Then he trained for the Presbyterian ministry at the University of Edinburgh. In 1733, he became a minister of the Presbyterian chapel in Tunbridge Wells, 35 miles southeast of London.

Thomas Bayes was a strong Newtonian in his scientific outlook. Thomas Bayes' early work appears to have been related mainly to infinite series, which was one of the paths followed by British mathematicians in the 18th century. Bayes' interest in probability has several origins. First, Bayes learned probability from Abraham de Moivre. Next, Bayes became interested in probability after reviewing a publication of Thomas Simpson, a special case of the law of large numbers: the mean of a set of observations is a better estimate of a location parameter than a single observation.

Bayes set out his theory of probability in "Essay Towards Solving a Problem in the Doctrine of Chances," published in the *Philosophical Transactions of the Royal Society of London* in 1764.

Bayes defined the problem as follows:

> Given the number of times in which an unknown event has happened and failed: Required the chance that the probability of its happening in a single trial lies somewhere between any two degrees of probability that can be named.

Bayes solved this problem by considering an experiment on a table (could have been a billiards table).

A ball is thrown across the table in such a way that it is equally likely to come to rest anywhere on the table. Through the point that it comes to rest on the table, draw a line. Then throw the ball *n* times and count the number of times it falls on either side of the line. These are the successes and failures. Under this physical model one can now find the chance that the probability of success is between two given numbers.

It was Bayes' friend Richard Price who communicated the paper to the Royal Society two years after Bayes' death in 1761. Bayes' fame rests on this result [3].

Bayes Theorem

What we have seen so far are called *classic probability theories* championed in the 17th century in France.

There is another system of probability, invented and advanced by Bayes in the 18th century in England (see Box 11.4 for a short biography).

In Bernoulli's system, the future is predicted by current probability derived from current data. In the Bayesian system of thinking, the probability of a future event is influenced by history too. Future probability is a product of current and historic probabilities. Extending it further, future probability is a product probability derived from data and theoretical probability derived from knowledge. Bayes boldly combined soft (subjective) and hard (derived from data) probabilities, a notion that remained unacceptable to many statisticians for years but widely adopted now. Bayes used the notion of conditional probability.

We can define conditional probability in terms of absolute probabilities: $P(A|B) = P(A \text{ and } B)/P(B)$; that is, the probability that A and B are both true divided by the probability that B is true.

Bayes used some special terms. Future probability is known as *posterior probability*. Historic probability is known as *prior probability*. Future probability can only be a likelihood, an expression of chance softer than the rigorous term probability. Future probability is a conditional probability.

A Clinical Lab Example

A simple illustration of the Bayes analysis is provided by Trevor Lohrbeer in *Bayesian Maths for Dummies* [4]. The gist of this analysis is as follows:

> A person tests positive in a lab. The lab has a reputation of 99% correct diagnosis but also has false alarm probability of 5%. There is a background information that the disease occurs in 1 in 1000 people (0.1% probability). Intuitively one would expect the probability that the person has the disease is 99%, based on the lab's reputation. Two other probabilities are working in this problem: a background probability of 0.1% and a false alarm probability of 5%. Bayes theorem allows us to combine all the three probabilities and predict the chance of the person having the disease as 1.94%. This is dramatically less than an intuitive guess.

The Bayesian breakthrough is in that general truth (or disease history) prevails upon fresh laboratory evidence. Data 11.1 presents the following three probabilities that define the situation.

P_1: probability of correct diagnosis
P_2: probability of false alarm
P_3: prevalent disease probability (background history)

Data 11.1 Bayes Estimation with Variable Disease Probability

Given Lab Characteristics		
P_1	Reputation of correct diagnosis	99%
P_2	False alarm probability	5%

Question: What is the probability P_0 of a person who tests positive having the disease?

P_3 Disease Probability %	P_0 Bayes Estimation Chance of Having Disease %
0.1	1.9
1.0	16.7
10.0	68.8
20.0	83.2
30.0	89.5
40.0	93.0
50.0	95.2
60.0	96.7
70.0	97.9
80.0	98.8
90.0	99.4

Note: It may be seen that posterior probability depends on prior probability.

The question is "What is the probability of a person who tests positive having the disease?" This probability is denoted by P_0 in Data 11.1. P_1 and P_2 are fixed, and P_3 is varied. The associated P_0 is calculated according to the Bayes theorem:

$$P_0 = \frac{P_1 P_3}{P_1 P_3 + (P_2(1 - P_3))} \tag{11.13}$$

In this formula, probabilities are expressed in fractions.

This is a way of understanding how the probability that a hypothesis is true is affected by a new piece of evidence. It is used to clarify the relationship between theory and evidence.

The role played by false alarm probability on estimation of P_0 can also be calculated in a similar way. By keeping the P_3 (disease history) constant in the above example, we can vary P_2, false alarm probability, and see the impact on estimation (see Data 11.2).

As false alarm probability P_2 decreases, the probability of the subject having disease P_0 increases, tending toward the probability of correct diagnosis P_1.

Data 11.2   Bayes Estimation with Variable False Alarm Probability

Constants		
P_3	0.001	Disease history
P_1	0.99	Probability of correct diagnosis

Variables	
P_2	False alarm probability
P_0	Probability of the subject having disease

Question: What is the probability P_0 of a person who tests positive having the disease?

Bayes Estimation	
P_2	P_0
0.05000	0.01943
0.01000	0.09016
0.00100	0.49774
0.00010	0.90834
0.00001	0.99001

The above example illustrates the application of conditional probability and how it can modify our judgment, for the better.

Application of Bayes Theorem in Software Development

Chulani et al. [5] applied the Bayes theorem to software development. The Bayes theorem is elegantly applied to software cost models.

The Bayesian approach provides a formal process by which a-priori expert judgment can be combined with sampling information (data) to produce a robust a-posteriori model

$$Posterior = Sample \times Prior$$

In the above equation "Posterior" refers to the posterior density function summarizing all the information. "Sample" refers to the sample information (or collected data) and is algebraically equivalent to the likelihood function. "Prior" refers to the prior information summarizing the expert judgment. In order to determine the Bayesian posterior mean and variance, we need to determine the mean and precision of the prior information and the sampling information.

Chulani et al. have used the Bayesian paradigm to calibrate the Cost Construction Model (COCOMO), combining expert judgment with empirical data. This illustration has great significance and holds great promise. This makes us think differently about data. In a Bayesian sense, data include an intuitive guess. The study of Chulani et al. proves that a healthy collaboration between empirical data and an intuitive guess, such as available in Bayes, is a practical solution to a hitherto unsolved problem.

Fenton [6] used the Bayesian belief networks (BBNs), a commendable expansion of the Bayesian paradigm, to predict software reliability.

Bibi et al. [7] applied BBNs as a software productivity estimation tool. They find that BBN is a promising method whose results can be confirmed intuitively. BBNs are easily interpreted, allow flexibility in the estimation, can support expert judgment, and can create models considering all the information that lay in a data set by including all productivity factors in the final model.

Wagner [8] used BBNs inside a framework of activity-based quality models in studying the problem of assessing and predicting the complex concept of software quality. He observes,

> The use of Bayesian networks opens many possibilities. Most interestingly, after building a large Bayesian network, a sensitivity analysis of that network can be performed. This can answer the practically very relevant question which of the factors are the most important ones. It would allow to reduce the measurement efforts significantly by concentrating on these most influential facts.

A Comparison of Application of the Four Distributions and Bayes Theorem

In the case of the binomial distribution, the trials are independent of one another. Trials are done with replacement.

The hypergeometric distribution arises when sampling is performed from a finite population without replacement, thus making trials dependent on one another.

In NBD, the number of trials is not fixed. Trials go until a specified number of successes are obtained.

The geometric distribution is a special case of NBD where trials are observed until the first success is achieved.

Bayes theorem provides a way to combine historical distribution with fresh evidence.

BOX 11.5 THE THEORY THAT WOULD NOT DIE

Sharon McGrayne's book, *The Theory That Would Not Die: How Bayes' Rule Cracked the Enigma Code, Hunted Down Russian Submarines, and Emerged Triumphant from Two Centuries of Controversy*, presents a history of Bayes' theorem. The following is an excerpt from the review of this book in http://www.lesswrong.com.

Bayes' system was Initial Belief + New Data → Improved Belief. Mathematicians were horrified to see something as whimsical as a guess play a role in rigorous mathematics; this problem of priors was insurmountable.

Pierre-Simon Laplace, a brilliant young mathematician, and the world's first Bayesian, came to believe that probability theory held the key, and he independently rediscovered Bayes' mechanism.

Joseph Bertrand was convinced that Bayes' theorem was the only way for artillery officers to correctly deal with a host of uncertainties about the enemies' location, air density, wind direction, and more.

Geologist Harold Jeffreys made Bayes' theorem useful for scientists, proposing it as an alternative to Fisher's *p*-values and significance tests, which depended on "imaginary repetitions."

For decades, Fisher and Jeffreys were the world's two greatest statisticians, traded blows over probability theory in scientific journals and in public. Fisher was louder and bolder, and frequentism was easier to use than Bayesianism. This marked a short lived decline of the Bayesian paradigm.

In 1983, the US Air Force sponsored a review of NASA's estimates of the probability of shuttle failure. NASA's estimate was 1 in 100,000. The contractor used Bayes and estimated the odds of rocket booster failure at 1 in 35. In 1986, *Challenger* exploded. Frequentist statistics worked okay when one hypothesis was a special case of another, but when hypotheses were competing and abrupt changes were in the data, frequentism did not work.

One challenge had always been that Bayesian statistical operations were harder to calculate, and computers were still quite slow. This changed in the 1990s, when computers became much faster and cheaper than before.

Review Questions

1. Give an example of the application of binomial distribution.
2. Give an example of the application of the hypergeometric distribution.
3. Give an example of the application of the negative binomial distribution.
4. Give an example of the application of the geometric distribution.
5. What is Bayes theorem?

Exercises

1. In a certain school, it has been estimated that the probability of students passing mathematics tests is 69%. Find out the probability of at least 80 passes in a batch of 89 students. Plot the related binomial distribution. Use Excel function BINOM.DIST for your calculations.
2. In a certain application 12 modules have just been built. Test results show that there are 4 defective modules. If we draw samples of size 3 without replacement, find the probability that a sample contains two defective modules. Use the Excel function HYPGEOM.DIST for your calculations.
3. Right first-time design probability in a software development project is estimated at 0.3. Estimate the probability of needing seven trials to find a defect-free feature design. Plot a graph between trials and geometric probability.

References

1. B. Singh, R. Viveros and D. L. Parnas, *Estimating Software Reliability Using Inverse Sampling*, Communications Research Laboratory, McMaster University, Hamilton Department of Mathematics and Statistics, The College of Information Sciences and Technology, The Pennsylvania State University. Available at http://citeseerx.ist.psu.edu/viewdoc/download?doi=10.1.1.71.1577&rep=rep1&type=pdf.
2. M. Wisal, *Formal Verification of Negative Binomial Distribution in Higher Order Logic*, Department of Computer Engineering, University of Engineering and Technology, Taxila, 2012.
3. D. R. Bellhouse, The Reverend Thomas Bayes, FRS: A biography to celebrate the tercentenary of his birth, *Statistical Science*, 19(1), 3–43, 2004.
4. T. Lohrbeer, *Bayesian Maths for Dummies*. Available at http://blog.fastfedora.com/2010/12/bayesian-math-for-dummies.html.
5. S. Chulani, B. Boehm and B. Steece, *Bayesian Analysis of Empirical Software Engineering Cost Models*, University of Southern California, IEEE Transactins on Software Engineering, USC-CSE, 1999.
6. N. Fenton, *Software Project and Quality Modelling Using Bayesian Networks*. Available at http://www.eecs.qmul.ac.uk/~norman/papers/software_project_quality.pdf.
7. S. Bibi, I. Stamelos and L. Angelis, *Bayesian Belief Networks as a Software Productivity Estimation Tool*, Department of Informatics, Aristotle University, Thessaloniki, Greece, 2003.
8. S. Wagner, A Bayesian network approach to assess and predict software quality using activity-based quality models. In: *Proceeding PROMISE '09 Proceedings of the 5th International Conference on Predictor Models in Software Engineering*, Article No. 6, ACM, New York, 2009.

Suggested Reading

Wroughton, J. and T. Cole, Distinguishing between binomial, hypergeometric and negative binomial distributions, *Journal of Statistics Education*, 21(1), 2013. Available at http://www.amstat.org/publications/jse/v21n1/wroughton.pdf.

Chapter 12

Law of Rare Events

Science pursues the study of rare events with fervor. The probability of rare events is a skewed one. In this chapter, we discuss one continuous distribution and one discrete distribution to represent rare events.

BOX 12.1   AGE DETERMINATION—CARBON DATING

In the mid-1940s, Willard Libby, then at the University of Chicago, realized that the decay of carbon 14 might lead to a method of dating organic matter. Wood samples taken from the tombs of two Egyptian kings, Zoser and Sneferu, were dated by radiocarbon measurement to an average of 2800 BC plus or minus 250 years. These measurements, published in *Science* in 1949, launched the "radiocarbon revolution" in archaeology and soon led to dramatic changes in scholarly chronologies. In 1960, Libby was awarded the Nobel Prize in chemistry for this work. The equation governing the decay of a radioactive isotope is

$$N = N_0 e^{-\lambda t}$$

where N_0 is the number of atoms of the isotope at time $t = 0$, N is the number of atoms left after time t, and λ is a constant that depends on the particular isotope. It is an exponential decay. Using this equation, the age of the sample can be determined. (http://en.wikipedia.org/wiki/Radiocarbon_dating)

Exponential Distribution

It is common knowledge in science that decay is defined as an exponential form, a simple and beautiful mathematical structure, as given in the following equation:

$$f(t) = e^{-\lambda t} \tag{12.1}$$

where λ is a constant describing the rate of decay and t is the time variable.

Nobel laureate Ernest Rutherford used this equation to describe the radioactive decay of thorium in 1907 [1].

> *It is the basis for the Nobel Prize in Chemistry he was awarded in 1908 "for his investigations into the disintegration of the elements, and the chemistry of radioactive substances."*

If we use the Geiger counter and counted the radiated particles, the data will fit a discrete Poisson distribution. If we measure loss of weight of the parent or the interarrival time of particles, the data will fit a continuous Exponential distribution. When we fit a curve to Rutherford data, we will obtain the following equation:

$$y = 144.03e^{-0.172x}$$
$$R^2 = 0.9991 \tag{12.2}$$

Figure 12.1 shows the Rutherford data and the exponential plot. The exponential form fits like a glove to the decay data. Rutherford also defined a parameter called *half-life*, the time taken for the parent matter to lose half its weight. On the exponential graph, half-life represents the median. The half-life point is marked in Figure 12.2 (4.03 days), where thorium activity becomes half of the start value. The start value is 144, and the half value is 72. The time required for this loss of activity is 4.03 days.

> *The exponential distribution is memoryless.*

This can be demonstrated using Figure 12.2. For thorium activity to drop from 72 to a half of 72, that is, 36, it will take another 4.03 days. This is exactly the time taken for thorium activity to drop from 144 to 72. The second drop takes the same time as the first drop because the exponential curve has no memory of the first drop. Each time, decay starts afresh with a new account and a fresh experience of the same half-life. Half-life is the property of the decaying matter represented in the exponential form. For thorium activity, it is 4.03 days.

The exponential nature of radioactive decay is exploited in carbon dating (see Box 12.1).

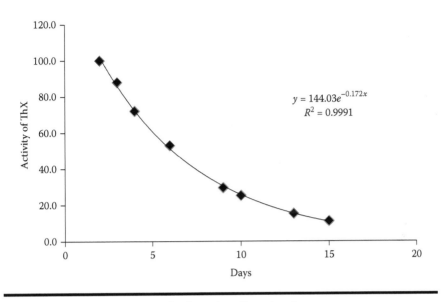

Figure 12.1   Exponential distribution of Radioactive decay.

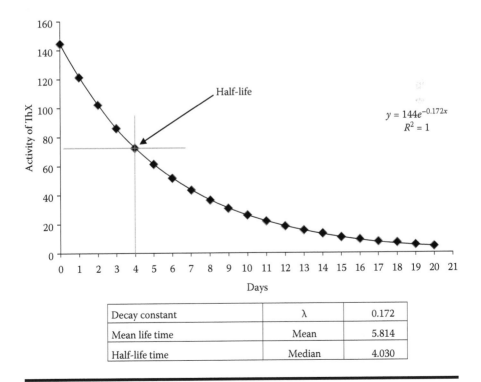

Decay constant	λ	0.172
Mean life time	Mean	5.814
Half-life time	Median	4.030

Figure 12.2   Half-life analysis using exponential distribution.

It can be seen that Equation 12.1 is just a variant of the proper exponential probability density function (PDF) shown as follows:

$$f(t) = \lambda e^{-\lambda t} \tag{12.3}$$

The cumulative distribution function (CDF) is as follows:

$$F(t) = 1 - e^{-\lambda t} \tag{12.4}$$

where t is time and λ is the rate constant. The mean of this distribution is $1/\lambda$. The standard deviation is also equal to $1/\lambda$.

In engineering, exponential distribution is primarily used in reliability applications. In the context of reliability, λ is known as the *failure rate* or *hazard rate*. In a chemical engineering example, corrosion rate is represented in exponential form. In an electrical engineering example, electrical charge stored in a capacitor decays exponentially. In a geophysics example, atmospheric pressure decreases exponentially with height.

Equation 12.3 shows that a single parameter completely specifies the PDF, a unique aspect responsible for the simplicity of the equation.

The other model statistics are as follows:

The median is $\dfrac{\ln 2}{\lambda}$.
The mode is 0.
The skewness is 2.
The kurtosis is 9.

The metric% software defects discovered during system testing decreases exponentially with time, as shown in Figure 12.3. Initial test effort discovers more defects, and subsequent tests begin to show lesser results, a common experience in software testing. We assume that risky modules are tested first, as per a well-designed test strategy. Representing defect metrics is a classic application of the exponential model.

Defects found in a testing day are counted and summed up to obtain Figure 12.3. The x-axis of the plot could be test day or even calendar day. We can plot total defects found every week and establish the exponential nature.

In reliability analysis, the median value $\dfrac{\ln 2}{\lambda}$ is called *half-life*. The mean $1/\lambda$ is known as *mean time to fail* (MTTF). Also, $f(t) = e^{-\lambda t}$ is known as *survival function* or *reliability function*. If the MTTF of a bulb is 400 hours, the corresponding $f(t)$ would define the reliability of the bulb. As time goes on, the reliability would decrease, notably after 400 hours, and the reliability of the bulb can be calculated directly from the following expression:

$$\text{Bulb reliability} = e^{-(t/400)}$$

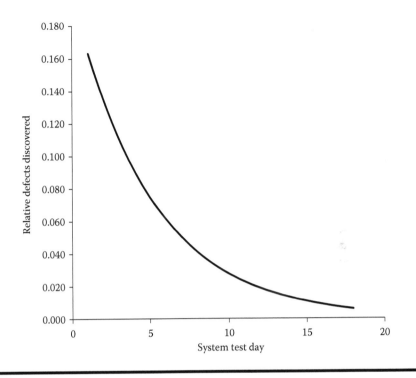

Figure 12.3   Exponential distribution of defect discovery.

where t is elapsed time in hours (see Box 12.2 for a note on a Super Bulb burning for 113 years).

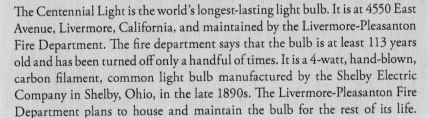

BOX 12.2   SUPER BULB

The Centennial Light is the world's longest-lasting light bulb. It is at 4550 East Avenue, Livermore, California, and maintained by the Livermore-Pleasanton Fire Department. The fire department says that the bulb is at least 113 years old and has been turned off only a handful of times. It is a 4-watt, hand-blown, carbon filament, common light bulb manufactured by the Shelby Electric Company in Shelby, Ohio, in the late 1890s. The Livermore-Pleasanton Fire Department plans to house and maintain the bulb for the rest of its life. (http://en.wikipedia.org/wiki/Livermore-Pleasanton_Fire_Department)

In software applications, there is a reversal of thinking: the CDF $F(t) = 1 - e^{-\lambda t}$ is known as the reliability function.

As time progresses, more defects are removed, and the product becomes more reliable, contrasting with the bulb. It is rather easy to calculate the fraction found until a given date by dividing the defects found until now by the total number of estimated defects, as shown in Figure 12.4. The cumulative defects found represent software reliability.

There is a caveat.

The rule says failure interval follows exponential distribution while defect events follow Poisson distribution. The essential truth is both follow the exponential law; one in continuous form, the other in discrete form.

In an ideal situation, we should use failure interval or time to fail in Equation 12.3 and plot a graph (that would resemble the same pattern in Figure 12.4). In real-life projects, the exact time of defect discovery is not always available. People accumulate information and submit reports on a weekly basis, occasionally on a daily basis, never on an hourly basis, unless of course if the bug tracking tool has a provision to capture defect events precisely in real time. Hence, we move away philosophically from reporting defect counts to reporting a metric called *defects per week*. Some people use defect density (defects per KLOC or defects per FP) instead of defect count. Either way, we have a density metric, which would still fit into a model represented in Figure 12.4.

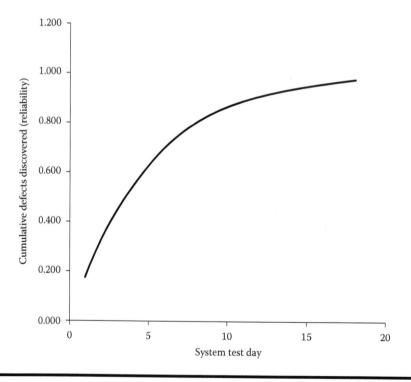

Figure 12.4 Exponential distribution cumulative distribution function (CDF) of cumulative defects found.

BOX 12.3 SIMÉON DENIS POISSON (1781–1840)

Siméon Denis Poisson was a French mathematician. His teachers Laplace and Lagrange quickly saw his mathematical talents. They became friends for life with their extremely able young student, and they gave him strong support in a variety of ways.

His paper on the theory of equations written in his third year was of such quality that Poisson could graduate without taking the final examination. He was employed as a tutor and appointed deputy professor 2 years later in 1802. In 1806, he became a full professor.

One of Poisson's contributions was the development of equations to analyze random events, later dubbed the Poisson distribution. It describes the probability that a random event will occur in a time or space interval under the conditions that the probability of the event occurring is very small but the number of trials is very large; hence, the event actually occurs a few times.

The fame of this distribution is often attributed to the following story. Many soldiers in the Prussian Army died due to kicks from horses. To determine whether this was due to a random occurrence or the wrath of god, the Czar commissioned a Russian mathematician to determine the statistical significance of the events. It was found that the data fitted remarkably well to a Poisson distribution. There was an order in the data, and deaths were now statistically predictable.

Poisson never tried experimental designs. He said,

> Life is good for only two things, discovering mathematics and teaching mathematics.

Poisson Distribution

The exponential law for discrete events can be expressed as follows:

$$P(x,\lambda) = \frac{e^{-\lambda}\lambda^x}{x!} \tag{12.5}$$

where x takes discrete integer values 0, 1, 2 ..., and λ is the mean value of x.

The Poisson distribution can be solved in Excel using the statistical function POISSON.DIST. For given values of x and λ, the function returns Poisson probability. While entering data by making cumulative = 0, we get probability distribution function, and by making cumulative = 1, we get cumulative probability.

Plots of Poisson probabilities of Equation 12.5 for λ = 1, 2, 3, and 4 are plotted in Figure 12.5.

As λ increases, the distribution shifts to the right and tends to turn symmetrical.

The corresponding cumulative probabilities are plotted in Figure 12.6. As λ increases, the curve attains an S shape.

The Poisson distribution was created by Siméon-Denis Poisson. In 1837, Poisson's Sorbonne lectures on probability and decision theory were published. They

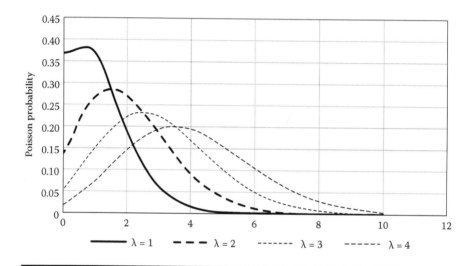

Figure 12.5 Poisson probability density function (PDF).

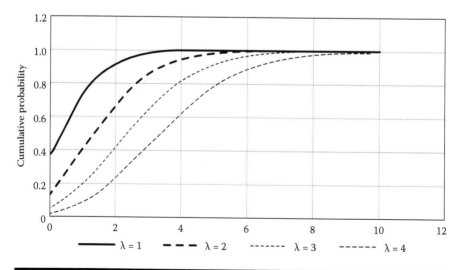

Figure 12.6 Poisson cumulative distribution function (CDF).

contained the Poisson distribution, which predicts the pattern in which random events of very low probability occur in the course of a very large number of trials. Poisson distribution is called the *law of rare events*.

A biographical note on the inventor of this distribution, Poisson, may be seen in Box 12.3. Poisson seems to have touched upon a universal law. Poisson distribution and its extensions are actively pursued by researchers in many domains, including software engineering.

A Historic Poisson Analysis: Deaths of Prussian Cavalrymen

In the historic data analysis done by von Bortkiewicz in 1898, deaths of Prussian cavalrymen due to horse kicks were fitted to a Poisson distribution. We can look at the data made available in *Statistics: The Poisson Distribution* [2], where the mean value of death per corps is given as $p = 0.5434$. Substituting this value in Equation 12.5 and treating x as the number of deaths, we can construct a Poisson distribution as follows:

$$P(x, 0.5434) = \frac{e^{-0.5434} 0.5434^x}{x!} \tag{12.6}$$

where x is number of deaths in a single corps.

Figure 12.7 shows a plot of this Poisson distribution.

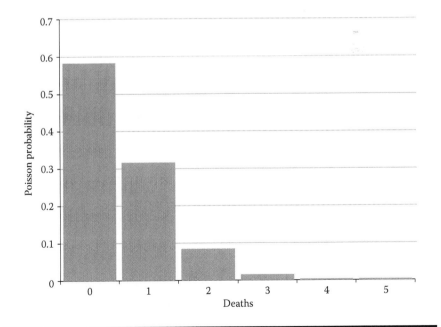

Figure 12.7 **Poisson distribution of Prussian cavalrymen deaths.**

Using the above Poisson distribution, Russian mathematician von Bortkiewicz predicted that "over the 200 years observed 109 years would be with zero deaths." It turned out that 109 is exactly the number of years in which the Prussian data recorded no deaths from horse kicks. The match between expected and actual values is not merely good, it is perfect.

BOX 12.4   ANALOGY—BAD APPLES

A truck delivering apples unloads at a warehouse. Most cartons have apples in good condition, but some apples are damaged. Typically, "damaged apples" is a rare event; only cartons in some part of the truck might be damaged. The occurrence of damaged apples is a Poisson process, the distribution of defects happens in spatial domain. The number of bad apples in unit volume is a Poisson parameter.

Likewise, a software product is shipped to the customer. When usage begins, some part of the product is found to have defects. Such defects are rare events. Across the code structure, defects are spatially distributed. However, software usage and defect discovery is a rare event in temporal domain. Hence, people use the word defect arrival rate. The number of defects arriving in unit time (e.g., a week) can be measured from defects counts in time. The defect arrival rate follows Poisson distribution.

Tests prior to release also discover defects in a similar manner. Defects "arrive" according to the Poisson distribution, in a broad sense. Change requests follow suit. Each development project has unique styles of managing defect discovery; accordingly, the Poisson distribution varies in structure and departs from the simple classic Poisson equation. There are several variants of the Poisson distribution to accommodate the different styles in defect management.

Analysis of Module Defects Based on Poisson Distribution

Before release, software defects are triggered by tests according to the Poisson distribution. Defect count in modules in User Acceptance Tests will be an example of rare events. If the average defects per module are 0.3 and if there are 100 modules in a release, the defects are distributed across the modules according to the Poisson distribution. All the modules are not likely to have equal defects. A few may have more and the count tapers off among the remaining. The distribution follows Equation 12.5. The plot of Poisson distribution is shown in Figure 12.8.

The mean of the distribution is now known as the *rate parameter*. The only parameter to the equation is the mean. Variance of the distribution is equal to mean. Hence, the statistical limits are known by simple formulas:

$$\text{UCL} = \lambda + 3\sqrt{\lambda} \qquad (12.7)$$

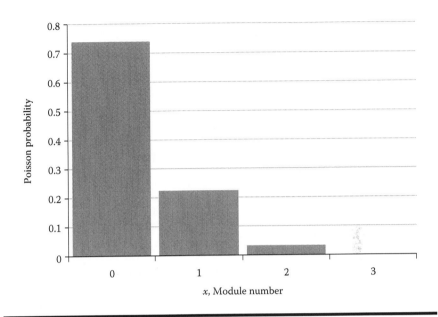

Figure 12.8 Poisson distribution of module defects.

$$CL = \lambda \tag{12.8}$$

$$LCL = \lambda - 3\sqrt{\lambda} \tag{12.9}$$

In the previous example,

$$\text{Upper control limit (UCL)} = 0.3 + 3\sqrt{0.3}$$
$$= 0.3 + 3 \times 0.548$$
$$= 1.943$$

This reasoning leads us to think of 1.943 defects per module as the statistical limit. Poisson approximation thus allows us statistical control of defects. Any module with more than 1.943 defects is a Poisson outlier. Poisson distribution here serves as a quality judge. (The use of this characteristic Poisson distribution is illustrated in Box 12.4.)

The CDF of the Poisson distribution, shown in Figure 12.9, is of special relevance to software defect management.

It clearly shows only a few modules contain defects. The rest have zero defects. This distribution helps to spot those defect intensive modules and subject them to appropriate testing.

Another help from the Poisson distribution study is an objective estimate of the right first-time index for the software product. This is the Poisson probability that

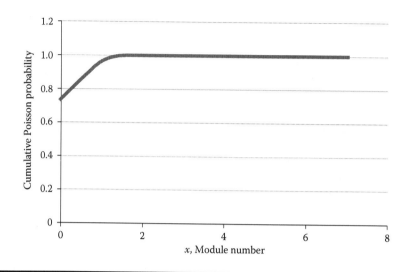

Figure 12.9 Cumulative Poisson probability of module defects.

zero modules will have defects. For example, in the product described in Figure 12.7, the Poisson probability is 0.74 for $x = 0$. This is the right first-time index.

Study of Customer Complaint Arrival Rate Based on Poisson Distribution

Customer complaints regarding field failure arrive at the Poisson rate. Let us consider a case where the average number of failure complaints arriving per month is λ. One question that comes to our mind is "Can we think about the maximum number of complaints that are likely to arrive?" Is there enough evidence in λ to predict the maximum number of complaints? Poisson distribution is applied to such cases. A c chart is plotted with the number of complaints arriving per month. The upper limit in the chart is calculated by the same formula used above. We find that the maximum number of complaints likely to arrive per month is

$$\lambda + 3\sqrt{\lambda}$$

This number could defy intuitive judgment of customer complaints; intuitive judgment hovers around the average value. The predicted number may exceed the maximum ever number of complaints received in any given month so far. The Poisson boundary easily exceeds the trend forecast. The Poisson approximation to customer complaint arrival is a very valuable aid.

An example of customer complaints arrival is shown in Figure 12.10.

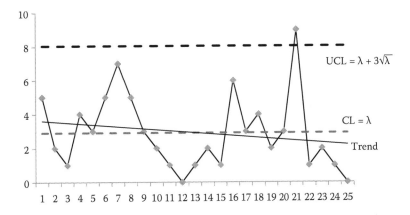

Figure 12.10   Control chart for customer complaints.

The Poisson boundary is marked as an upper bound. A trend line is also included in the figure to show how the Poisson boundary shows a complaint rate higher than indicated by the trend.

In this context, the application of Poisson distribution to model baseball events is illustrated in Box 12.5.

Applying Poisson Distribution to Software Maintenance

The arrival of service requests follows Poisson distribution. The interarrival time follows exponential distribution. Both are memoryless. The time to repair a bug does not depend on previous records. Mean time to fix bugs in a particular setup controls dispersion of results. Month or week, teams may experience the same Poisson curves. Sophisticated models for queues have been built, but the building block is the exponential law.

Bathtub Curve of Reliability: A Universal Model of Rare Events

Failure of components is extensively used in reliability analysis because it is a Poisson process. The bath tub curve of reliability has three zones. The first is characterized by a rapidly decreasing failure rate. This region is known as the *early failure period* (also called *infant mortality period*). Next, the failure rate levels off and remains constant in the flat portion of the bathtub curve. Finally, the failure rate begins to increase as materials wear out and degradation failures occur at an ever increasing rate. This is the *wear out failure period* (see Figure 12.11).

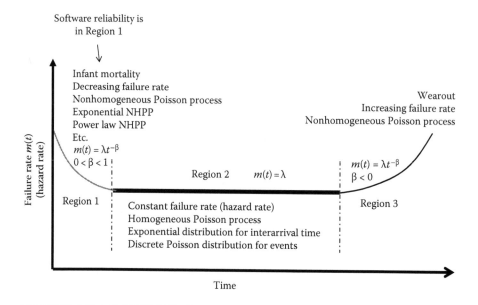

Figure 12.11   Bath tub curve.

The flat bottom is governed by a Poisson process that has a constant failure rate (or hazard rate). To be more specific, this is called as the *homogeneous Poisson process* (HPP). The term homogeneous is due to a constant failure rate or hazard rate. The infant mortality period is also a period of growth in reliability. The failure rate in this period is not constant but steadily reducing; hence, it is called *nonhomogeneous*. This associated process is the nonhomogeneous Poisson process (NHPP). The wear out period is also an NHPP, the difference being the fact that failure rate here steadily increases until the system is discarded.

It may be noted that in the example of the bath tub curve shown in Figure 12.11,

■ In the infant mortality period,

$$m(t) = \lambda t^{-\beta}, 0 < \beta < 1 \qquad (12.10)$$

defines a decreasing failure rate and an NHPP.

■ In the middle region,

$$m(t) = \lambda, \beta = 0 \qquad (12.11)$$

defines a constant failure rate and an HPP.

■ In the wear out period

$$m(t) = \lambda t^{-\beta}, \beta < 0 \qquad (12.12)$$

defines an increasing failure rate and an NHPP, in a reverse direction.

The HPP can be described only by exponential law, an oversimplification though. Constant failure rate, a key assumption in HPP, is too ideal to be true even in the case of mechanical systems. A bulb, under HPP, will have the same reliability after burning through 400 hours or any time fixed by the analyst. Physically this is meaningless. Similarly, the physical meaning of a failure rate in a situation shown in Figure 12.2 begs explanation. No software ever operates at a constant failure rate, although the exponential representation produces such a parameter. It must be borne in mind that Figure 12.2 has been obtained by numerical curve fitting rather than by using physically reasonable reliability parameters such as failure rates or MTBF or MTTF.

The bath tub curve, in its entirety, is true for mechanical systems. In the case of software, failures are constrained to Region 1, which records reliability growth. Hence, software failure models are called *reliability growth models*. For both the cases, we now need the help of NHPP modeling for a more accurate representation of real world failure patterns.

Nonhomogeneous Poisson Process (NHPP)

Real-life software defect arrival is more complex than simple exponential curves, an example available in Figure 12.12. It presents a typical defect arrival pattern during system testing. Approximately 140 defects are discovered over a time span of about three months. It is not a smooth exponential cumulative distribution. The curve is

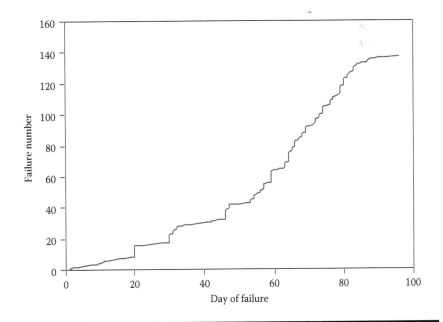

Figure 12.12   Defect arrival pattern—empirical model.

irregular; defects seem to be triggered in different rates in different periods. There seem to be short test cycles within the main testing process. The curve also has several linear climbs. These details mark clear departure from the simple exponential, making a case for building NHPP.

The equation to an NHPP is given as follows:

$$P(x, m(t)) = \frac{e^{-m(t)} m(t)^x}{x!} \qquad (12.13)$$

where $m(t)$ is the mean value function that takes the place of the traditional failure rate constant λ in the HPP model presented in Equation 12.5.

It may be noted that Equation 12.13 is exactly Equation 12.5, except for a redefinition of the rate constant λ. An NHPP is completely defined by its mean value function $m(t)$. Building an NHPP model decreases the identification of the right function for $m(t)$ and the derivation of the parameters of the function from failure data.

There are many options available to choose a function for $m(t)$. Researchers have used different functions to suit different situations. The list includes exponential, logarithms, Gaussians, Weibulls, and logistic functions. Even mixtures of functions have been used to deal with complex events.

It is now a custom to think of NHPP with two equations. The bigger Poisson equation in Equation 12.13 defines the structure, and the mean value function in Equations 12.10–12.13 defines a central component. In fitting NHPP to data, we derive the coefficients of the mean value function $m(t)$ from data. There is no need to consult the Poisson equation for this purpose. The Poisson is in the background, as an abstraction of the model.

Think of NHPP, think of mean value function.

An early application of the NHPP power law is by Duane [3], who in 1964 observed,

When he plotted cumulative MTBF estimates versus the times of failure on log-log paper, the points tended to line up following a straight line. This was true for many different sets of reliability improvement data and many other engineers have seen similar results over the last three decades. This type of plot is called a Duane Plot and the slope beta of the best line through the points is called the reliability growth slope or Duane plot slope. A straight line on a Duane plot is equivalent to the NHPP Power Law Model.

The NHPP power law has been used as a model for "reliability improvement."

Goel–Okumoto (GO-NHPP) Model

Many models have been proposed by the studies about software reliability based on NHPP. The main reason for selecting the NHPP technique is that it facilitates the testers an analytical framework that helps in identifying the faults derived from the software during the testing process, as noted by Vamsidhar et al. [4].

One of the most widely models used is the Goel–Okumoto NHPP model. On the basis of their study of actual failure data from many systems, Goel–Okumoto proposed the following exponential mean value function for their NHPP model [5]:

$$m(t) = a(1 - e^{-bt}) \qquad (12.14)$$

where $m(t)$ is the expected cumulative number of defects function, a is the expected total number of defects in the system, and b is the defect detection rate per defect.

The model assumes exponential behavior of failure and perfect debugging so that failure intensity reduces with time. It is exactly a replica of the cumulative exponential distribution given in Equation 12.4. The familiar λ, the failure rate constant, is now called b, the detection rate constant, supporting the paradigm that failure events in software are detection events. b could also stand for test case efficiency. The constant a is a scaling factor, introduced to represent the number of defects in the product. The Goel–Okumato model treats a as a parameter to be estimated from data.

It may be noted that the Goel–Okumuto model fits data to the exponential distribution. This strengthens the application of the exponential law in software reliability engineering. What is interesting is even in NHPP, the exponential law prevails as a fundamental principle. This upholds a universal view: "the exponential function is used to generate several other functions."

By substituting the Goel–Okumoto mean value function in the NHPP equation, we get the following detailed expression of NHPP:

$$P(x, m(t)) = \frac{e^{-a(1-e^{-bt})} (a(1 - e^{-bt}))^x}{x!} \qquad (12.15)$$

where x takes discrete integer values 0, 1, 2, and so on.

The detailed expression still has only two parameters, a and b. Applying Equation 12.9 for any given time t, we can create the Poisson probabilities of finding x number of defects. Plotting an NHPP is a complex thing to do. We have plotted the mean value function in Figure 12.13. Alongside, we have also plotted cumulative NHPP probabilities for the following discrete x values, for example, $x = 0$, $x = 1$, $x = 2$, and $x = 3$. This creates a family of curves for Equation 12.9.

Fitting the mean value function to data is the real job in building an NHPP—a curve-fitting job, an empirical task.

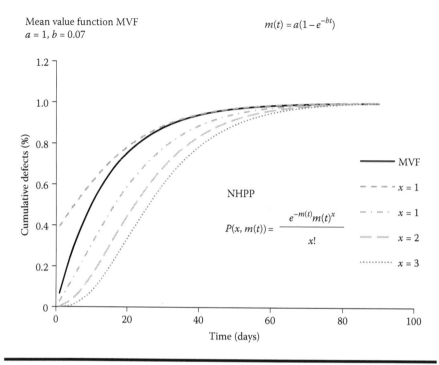

Figure 12.13 Goel–Okumoto model.

> God does not care about our mathematical difficulties.
> He integrates empirically.
>
> **Albert Einstein**

There are several references to the use of the Goel–Okumoto model, often called the GO model. A few are mentioned in the following section.

Different Applications of Goel–Okumoto (GO) Model

The law of rare events is fully realised in structure of GO-NHPP model, as we have seen. The GO NHPP Model has been extensively researched and used as a Software Reliability Growth Model (SRGM). A few attempts are listed below. The wide varieties of applications of the GO model explore the model features and identify the limits.

1. Nagar and Thankachan [6] have used the GO model to "decide the amount of more testing required and for the correct estimation of the remaining errors." They call b as a roundness factor, similar to a shape factor that tends to zero when irregularity increases.

2. Wood [7] considers nine SRGMs and has included the GO model in his study. He draws special attention to the collection of time data, the argument in SRGMs. He also proposes a two-stage NHPP "if a significant amount of new code is added during the test period."

3. In a survey of software reliability models, Pai [8] considers the GO model. He makes a salient observation regarding the GO model usage, "The model requires failure counts in the testing intervals and completion time for each test period for parameter estimation."

   In our opinion, this practice gives extra credibility to the mean value function and invests it with more decision making power. An exponential model that uses sums of the defects found in test intervals and the completion time of the test interval as the argument does not depend so much on Poisson abstraction.

4. Liu et al. [9] propose a generalized NHPP that uses a bell curve for fault detection rate. The bell curve handles variations due to fluctuations in debugging, learning, and fault removal efficiency. The results show that the proposed model fits failure data better than some selected NHPP models, including the GO model.

5. Anjum et al. [10] have evaluated 16 SRGMs proposed during the past 30 years using a set of 12 comparison criteria. They find the GO model in position 6 from the top. Surprisingly, they find the generalized Goel model in the 14th position.

   The generalized Goel model does not use the simple exponential law for its mean value function but adds a third parameter to generate desired shape changes. A graph of the generalized Goel model is available in Chapter 21.

6. Mohd and Nazir [11] have studied different reliability models and find an interesting characteristic in the GO model.

   It should be noted that here the number of faults to be detected is treated as a random variable whose observed value depends on the test and other environmental factors. This is a fundamental departure from the other models which treat the number of faults to be a fixed unknown constant.

7. Kim et al. [12] find that the GO model can be applied to safety critical software, although

   it is generally known that software reliability growth models such as the Jelinski-Moranda model and the Goel–Okumoto's non-homogeneous Poisson process (NHPP) model cannot be applied to safety-critical software due to a lack of software failure data.

   Their analysis confirms the fear: the estimated total number of inherent software faults varies from 27.32 to 34.83 for the GO model as the software failure numbers change from 24 to 34. Results are sensitive to the number of failure data points.

8. Lin and Huang [13] finds the Weibull model better than the GO model in a special application but chooses to refer to the GO model as a benchmark.

9. Gokhale and Trivedi [14] propose an enhanced NHPP, called the mean value function, as a coverage function and use the log logistic function instead of exponential to get better results than the GO model.

BOX 12.5   RARE BASEBALL EVENTS

Huber and Glen [15] have studied three sets of rare baseball events—pitching a no-hit game, hitting for the cycle, and turning a triple play—which offer excellent examples of events whose occurrence may be modeled as Poisson processes. From 1901 to 2004, there have been 206 no-hitters, 225 cycles, and 511 triple plays. The associated mean values per year have been calculated as follows:

No-hitter = 1.98.
Cycle = 2.16.
Triple plays = 4.91.

The above mean values characterize the respective Poisson distributions. The researchers have also calculated mean interarrival times as follows:

No-hitter = 772 games.
Cycle = 720 games.
Triple play = 316 games.

Using the mean values, intertribal times have been fitted to exponential distributions. The researchers find a good fit between the actual data and exponential fit, except in the case of triples.

Overall, this is a very good illustration of building Poisson and exponential models for rare events.

Review Questions

1. Relate the Poisson distribution to the exponential distribution.
2. How does the nonhomogeneous Poisson process (NHPP) differ from homogeneous Poisson process (HPP)?
3. Why is the exponential distribution considered as a fundamental engineering curve?
4. How does carbon dating illustrate a fundamental application of the exponential distribution followed by nature?
5. How did the Prussian cavalrymen death data prove that the Poisson distribution works?

Exercises

1. Applying the Goel–Okumoto NHPP model defined in Equation 12.8, given that $b = 0.04$, estimate time t to reach a reliability level of 0.95. Let us denote this time as t_{95}.
2. Apply Equation 12.9 and find out the probability of finding two defects at a point of time $= t_{95}$. Clue: substitute $x = 2$ in Equation 12.9. Also consult Figure 12.13 for understanding the problem.
3. If the average defects per module $= 0.4$, find the right first-time index of the application.
   Clue 1: RFT is the probability of getting zero defects in a module during testing.
   Clue 2: Use Excel function POISSON.DIST to calculate this number.
4. Let us take the example of testing 100 components in an application. The average defect per module is 0.2. What is the upper control limit on a quality control chart for the components?
5. Assume the Power Law for NHPP. The constant $b = 0.5$. The failure rate of an application is 5 defects per week immediately after release. What would be the failure rate in the fifth week?

References

1. E. Rutherford and F. Soddy, The cause and nature of radioactivity: Part I, *Philosophical Magazine*, 4, 370–396 (1902) (as reprinted in *The Collected Papers of Lord Rutherford of Nelson*, 1, pp. 472–494 [London: Allen and Unwin, Ltd., 1962]).
2. L. J. Bortkiewicz, The Poisson distribution, *The Law of Small Numbers*, Statistics, 1898. Available at http://en.wikipedia.org/wiki/Poisson_distribution.
3. J. T. Duane, *Power Law NHPP*, 1964. Available at http://www.itl.nist.gov/div898/handbook/apr/section1/apr191.htm.
4. Y. Vamsidhar, Y. Srinivas and A. Brahmani Devi, A novel methodology for software reliability using mixture models and non-homogeneous Poisson process, *International Journal of Advanced Research in Computer Science and Software Engineering*, 3(6), 2013.
5. S. H. Kan, *Metrics and Models in Software Quality Engineering*, 2nd ed., Addison-Wesley Professional, 2002.
6. P. Nagar and B. Thankachan, Application of Goel–Okumoto model in software reliability measurement, *International Journal of Computer Applications (0975–8887) on Issues and Challenges in Networking, Intelligence and Computing Technologies*, 2012.
7. A. Wood, *Software Reliability Growth Models*, Tandem computers, 1996.
8. G. Pai, *A Survey of Software Reliability Models*, Department of Virginia, VA, 2002.
9. H.-W. Liu, X.-Z. Yang, F. Qu and Y.-J. Shu, A general NHPP software reliability growth model with fault removal efficiency, *Iranian Journal of Electrical and Computer Engineering*, 4(2), Summer–Fall, 2005.
10. M. Anjum, A. Haque and N. Ahmad, Analysis and ranking of software reliability models based on weighted criteria value, *International Journal of Information Technology and Computer Science*, 5(2), 2013.
11. R. Mohd and M. Nazir, Software reliability growth models: Overview and applications, *Journal of Emerging Trends in Computing and Information Sciences*, 3(9), 2012.
12. M. C. Kim, S. C. Jang and J. J. Ha, *Possibilities and Limitations of Applying Software Reliability Growth Models to Safety-Critical Software*, Korea Atomic Energy Research Institute, 2006.
13. C.-T. Lin and C.-Y. Huang, *Software Reliability Modeling with Weibull-Type Testing-Effort and Multiple Change-Points*, DCS, National Tsing Hua University, Taiwan.
14. S. S. Gokhale and K. S. Trivedi, A time/structure based software reliability model, *Annals of Software Engineering*, 8, 85–121, 1999.
15. M. Huber and A. Glen, Modeling rare baseball events—Are they memoryless? *Journal of Statistics Education*, 15(1), 2007.

Suggested Readings

Arekar, K., NHPP and S-shaped models for testing the software failure process, *International Journal of Latest Trends in Computing*, 1(2), 2010.

Barraza, N. R., Parameter estimation for the compound Poisson software reliability model, *International Journal of Software Engineering and Its Applications*, 7(1), 137–148, 2013.

Goel, A., Software reliability models: Assumptions, limitations and applicability, *IEEE Transactions on Software Engineering*, 11(12), 1411–1423, 1985.

Grottke, M., *Software Reliability Model Study*, IST-1999-55017, January 2001. http://www.grottke.de/documents/SRModelStudy.pdf.

Huan, C.-Y., M. R. Lyu and S.-Y. Kuo, A unified scheme of some nonhomogenous Poisson process models for software reliability estimation, *IEEE Transactions on Software Engineering*, 29(3), 261–269, 2005.

Huang, C.-Y., M. R. Lyu and S.-Y. Kuo, Unified scheme of some non-homogeneous Poisson process models for software reliability estimation, *IEEE Transactions on Software Engineering*, 29(3), 2003.

Jain, M. and K. Priya, Reliability analysis of a software with non homogeneous Poisson process (NHPP) failure intensity, *International Journal of Operational Research Society of Nepal*, 1, 1–19, 2012.

James, A. et al., Identification of defect prone classes in telecommunication software systems using design metrics, *Information Sciences*, 2006.

Kapur, P. K., K. S. Kumar, J. Prashant and S. Ompal, Incorporating concept of two types of imperfect debugging for developing flexible software reliability growth model in distributed development environment, *Journal of Technology and Engineering Sciences*, 1(1), 2009.

Khatri, S. K., P. Trivedi, S. Kant and N. Dembla, Using artificial neural-networks in stochastic differential equations based software reliability growth modeling, *Journal of Software Engineering and Applications*, 4, 596–601, 2011.

Massey, W. A., G. A. Parker and W. Whitt, Estimating the parameters of non homogeneous Poisson process with linear rate, *Telecommunication Systems*, 5, 361–388, 1996.

Meyfroyt, P. H. A., *Parameter Estimation for Software Reliability Models*, July 20, 2012. http://alexandria.tue.nl/extra1/afstversl/wsk-i/meyfroyt2012.pdf.

Nikora, A. P. *Software System Defect Content Prediction from Development Process and Product Characteristics*, A dissertation presented to the Faculty of the Graduate School University of California, May 1998. http://csse.usc.edu/csse/TECHRPTS/PhD_Dissertations/files/Nikora_Dissertation.pdf.

Purnaiah, B., Fault removal efficiency in software reliability growth model, *Advances in Computational Research*, 4(1), 74–77, 2012.

Qian, Z., Software reliability modeling with testing-effort function and imperfect debugging, *Telkomnika*, 10(8), 1992–1998, 2012.

Rafi, S. M., K. N. Rao and S. Akthar, Software reliability growth model with logistic-exponential test-effort function and analysis of software release policy, *International Journal on Computer Science and Engineering*, 2(2), 387–399, 2010.

Satya Prasad, R., V. Goutham and N. Pawan Kumar, Estimation of failure count data using confidence interval, *International Journal of Innovative Technology and Exploring Engineering*, 2(6), 2013. http://www.ijitee.org/attachments/File/v2i6/F0834052613.pdf.

Savanur, S., *Use Software Reliability Growth Model (SRGM) for Residual Defects Estimation*, Blog, January 1, 2013.

Shatnawi, O., Discrete time NHPP models for software reliability growth phenomenon, *International Arab Journal of Information Technology*, 6(2), 124, 2009.

Son, H. S., H. G. Kang and S. C. Chang, Procedure for application of software reliability growth models to NPP PSA, *Journal of Nuclear Engineering and Technology*, 41(8), 1065–1072, 2009.

Vamsidhar, Y., P. Samba Siva Raju and T. Ravi Kuma, Performance analysis of reliability growth models using supervised learning technique, *International Journal of Scientific and Technology Research*, 1(1), 2012.

Williams, P., Prediction capability analysis of two and three parameters software reliability growth models, *Information Technology Journal*, 5(6), 1048–1052, 2006.

Chapter 13

Grand Social Law: The Bell Curve

Most of us have been initiated into statistical thinking through normal distribution, with its well-known bell-shaped curve. The normal distribution was invented from the binomial distribution.

 The binomial distribution is discrete, the normal distribution is continuous.

de Moivre invented normal distribution in 1756. It is also called the *Gaussian distribution* because Gauss was the first to apply this equation (1809). Popularly, this distribution is known simply as the *bell curve* (see Box 13.1 for a brief history). This is widely used in science, engineering, economics, management, and a host of disciplines.

The basic form of the normal distribution, known as the *standard normal curve*, is defined in Equation 13.1, and the graph is shown in Figure 13.1.

$$y = \frac{1}{\sqrt{2\pi}} e^{-\frac{z^2}{2}} \tag{13.1}$$

The distribution peaks at the mean, is symmetric, and spreads from $-\infty$ to $+\infty$.

The equation for normal distribution is shown in Equation 13.2. It is defined by two parameters, mean μ and standard deviation σ. The mean is known as the location parameter because it controls the location of the distribution. The standard deviation is known as the scale parameter because it controls the scale (width) of

BOX 13.1 ORIGINS OF A SOCIAL LAW

Normal distribution has cast its influence in almost every field of life and research. It has gained the status of a social law.

French-born British mathematician Abraham de Moivre (1667–1754) published *A Doctrine of Chance: A Method of Calculating the Probabilities of Events in Play* in 1718, wherein he addressed the gambling problem. The third edition appeared in 1756; it contained the approximation to the binomial distribution by the normal distribution.

> de Moivre actually had written the equation down in 1708; obtained it as a limit of coins tossing or binomial distribution. We think of a coin being tossed 'n' times, and note the proportion of k heads. After many k-fold trials, we obtain a graph showing the number of occasions on which we get 0 heads, 1 head, 2 heads, ... n heads. The curve will peak around the probability of getting heads with the coin. As the number of tosses 'n' grows without a bound, a normal distribution results [1].

de Moivre's concern was with games of chance, and his discovery showed the power of sampling to determine patterns in a population by examining only a few members. He spent the last part of his life by solving problems of chance for gamblers as the resident statistician of Slaughter's Coffee House in London.

In 1809, German mathematician and astronomer Johann Carl Friedrich Gauss (1777–1855) showed that errors of measurement made in astronomical observations followed a symmetric distribution called *normal distribution*. Gauss was also the first to develop the utility of the normal distribution curve, which had been discovered earlier by de Moivre. This distribution is now often called *Gaussian*.

> The curve was developed by observational astronomers who used the ideas of normal distribution to verify the accuracy of measurements. They measured a distance many times and graphed the results. If most measurements clustered around the mean, then the average of the results could be considered reliable. Outliers or deviant measurements could be discounted as inaccurate [2].

The normal distribution has been studied under various names for nearly 300 years. To the historically inclined, it is Laplace's second law, Gaussian law, or Laplace–Gaussian curve. The names *law of deviation* and *error curve* could make more sense to experimenters. Pearson, Fisher, and Galton have called it the normal curve, the name greatly favored by statisticians.

Today, in statistics books, we tend to call this the normal distribution. In the world of science, the favored name is Gaussian distribution.

the distribution. There is no separate shape parameter because the shape is fixed: it is a bell shape.

$$F(x,\mu,\sigma) = \frac{1}{\sqrt{2\pi}\sigma} e^{-\frac{(x-\mu)^2}{2\sigma^2}} \qquad (13.2)$$

where μ is the mean (location parameter), and σ is the standard deviation (scale parameter).

Mean and standard deviation are part of descriptive statistics, discussed in Chapter 1. For any data set, we can estimate these two parameters. The equation is a natural sequel.

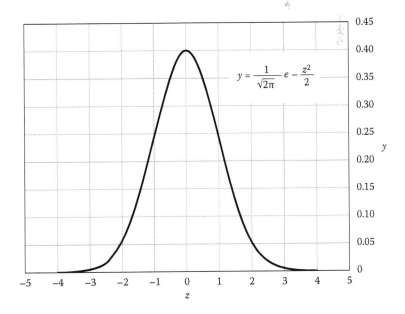

$$y = \frac{1}{\sqrt{2\pi}} e^{-\frac{z^2}{2}}$$

Figure 13.1   Standard normal curve.

The statistical properties of this distribution are as follows:

Mean = μ
Mode = μ
Median = μ
Kurtosis = 3
Relative kurtosis = 0
Skew = 0
Variance = σ^2
Standard deviation = σ^2
Range = $-\infty$ to $+\infty$

The mean code productivity in LOC per person-day and its standard deviation can be easily calculated from data and the corresponding normal distribution graph can be plotted.

In Figure 13.2, the assumed normal distribution of productivity is plotted for four different standard deviations. We have to assume normal distribution because productivity data would be seen as nonnormal had we plotted a histogram. However, we proceed with normal approximation. If dispersion decreases, it is a good sign; it indicates that the process becomes better. Figure 13.2 shows that as the standard deviation decreases, the height of the curve increases while its width decreases.

Real-world process improvement consists of reduction in variation and a simultaneous favorable shift in the mean. Figure 13.3 shows the bell curves for productivity improvement.

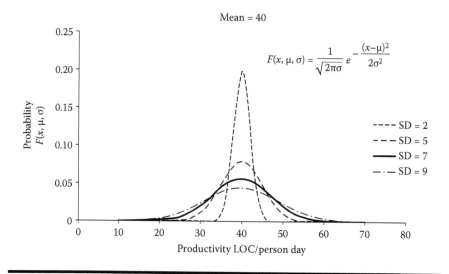

Figure 13.2 Gaussian probability density function (PDF) of productivity.

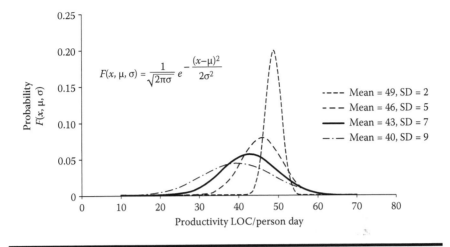

Figure 13.3   Gaussian model for productivity improvement.

The best performance is where the mean is 49 and the standard deviation 2. This gets closer to the oft spoken about rule of thumb of 50 LOC per person-day. The curves are still hypothetical, at best approximate. The bell curves in Figure 13.3 portray a story of improvement captured from a Gaussian lens.

First-Order Approximation of Variation

> If that enabled us to predict the succeeding situation with the same approximation, that is all we require, and we should say that the phenomenon had been predicted, that it is governed by the laws.
>
> **Henri Poincare**

Building a Gaussian is rather easy, from just two parameters, mean and standard deviation. These two can be obtained by expert judgment as well if data were not accessible. If we can guess optimistic and pessimistic values, we can "estimate" the Gaussian mean and standard deviation. The difference between the maximum and the minimum values is the estimated range. The rule of thumb we use to find standard deviation is given as follows:

$$\text{Standard deviation} = \frac{\text{Range}}{6} \qquad (13.3)$$

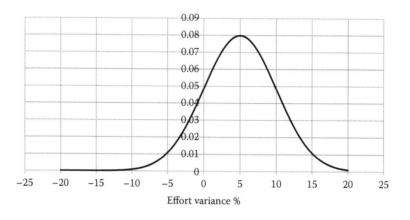

Figure 13.4 Gaussian distribution of effort variance.

The mean is recalled from the central tendency of the process. The ability of the Gaussian distribution to connect easily with approximate data makes it a "social law."

The Gaussian distribution allows us to see the two sides of truth: tendency and dispersion and facilitates fair judgment.

For example, from remembered mean value of effort variance = 5% and range = 30%, we can construct the Gaussian distribution shown in Figure 13.4. The central tendency, pictorially seen, reveals the problem. If planning and estimation practices were perfect, the central tendency would be zero. Nonzero tendency is a remark on project management.

BOX 13.2 IS THE BELL CURVE FAIR?

The power of the bell curve is linked to the central limit theorem (CLT): sample means tend to be normally distributed as sample size N tends to be large.

French mathematician Pierre-Simon Laplace rescued the CLT from the nearly forgotten work of Abraham de Moivre and published it in his monumental work *Théorie Analytique des Probabilités*. In 1901, Russian mathematician Aleksandr Lyapunov defined it in general terms and proved precisely how it worked mathematically [3].

Sir Francis Galton described the CLT as follows [4]:

> I know of scarcely anything so apt to impress the imagination as the wonderful form of cosmic order expressed by the "Law of Frequency of Error." Whenever a large sample of chaotic elements are taken in hand and marshaled in the order of their magnitude, an unsuspected and most beautiful form of regularity proves to have been latent all along.

The actual term *central limit theorem* was first used by George Pólya in 1920 in the title of a paper. Pólya referred to the theorem as central because of its importance in probability theory [5].

According to Le Cam, the French school of probability interprets the word central in the sense that "it describes the behaviour of the centre of the distribution as opposed to its tails" [6].

Between 1870 and 1913, Markov, Chebyshev, and Lyapunov contributed to CLT. During 1920 to 1937, Lindeberg, Feller, and Lévy perfected the CLT [7].

CLT sets the context for a bell curve paradigm. The science of measurements presents another truth. Whatever we measure, we make repetitions to make measurements credible, and we measure the bell curve of the measured parameter. The limit or peak of the bell curve is the truth. The tails denote errors. Criticism of the bell curve as a grading curve (by some educationalists) is ill founded. The bell curve represents data and cannot be made responsible for hypothesis.

To sum it up, the bell curve represents truth better than isolated data.

Estimation Error

Variance metrics is a double-edged sword. On the one side, it measures how well a plan is executed; on the other side, it measures how well a project is estimated. If applied to estimation, this metric can be renamed as the percentage of estimation error.

When processes mature, estimation errors tend to be the curve shown in Figure 13.5. Estimation errors are measurement errors; they resemble the astronomical measurement errors used by Gauss when he discovered a path breaking application of normal distribution. This is true in the case of size estimation, schedule estimation, effort estimation and even defect count estimation; errors in all these are Gaussian.

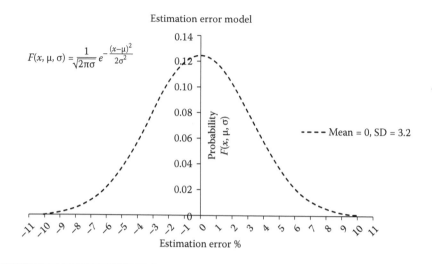

Estimation error model

$$F(x, \mu, \sigma) = \frac{1}{\sqrt{2\pi\sigma}} e^{-\frac{(x-\mu)^2}{2\sigma^2}}$$

- - - Mean = 0, SD = 3.2

Probability $F(x, \mu, \sigma)$

Estimation error %

Figure 13.5 Gaussian distribution of estimation errors.

Viewing Requirement Volatility

In the beginning of a project, managers do consider a risk of scope creep and plan out strategies to handle risk. There may not be objective evidence for potential scope creep, but approximate models based on benchmark data can be used to construct a Gaussian model to guide strategic planning. In certain projects, requirement volatility is believed to have a standard deviation of approximately 3.3% and a mean value of 4%, as a rule of thumb.

Thumb rule is merely an expression of one's experience.

With practice on statistical thinking, we can easily convert knowledge into Gaussian parameters.

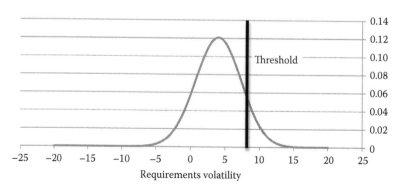

Threshold

Requirements volatility

Figure 13.6 Gaussian distribution of requirements volatility.

A plot of the Gaussian version of the above rule of thumb is shown in Figure 13.6. The tolerance limit is marked on the graph.

Figure 13.6 is a pictorial model to understand the challenge of requirement volatility in the context of a constraining limit. It provides a great visualization of a process along with process constraint (or goal).

Traditional ways to deal with information—reading, listening, writing, talking—are painfully slow in comparison to "viewing the big picture." Those who survive information overload will be those who search for information with broadband thinking but apply it with a single-minded focus.

Kathryn Alesandrini
Survive Information Overload: The 7 Best Ways to Manage Your Workload by Seeing the Big Picture

Risk Measurement

We can use the Gaussian curve to measure risk, and this is often carried out in software project management. For example, in Figure 13.6, the tolerance limit marks off a tail whose area indicates risk. In Figure 13.7, we show the Gaussian with the tail area marked in black.

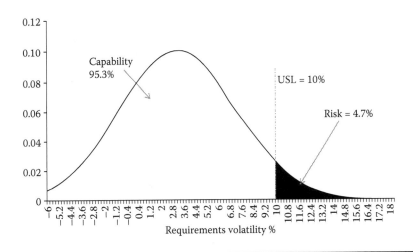

Figure 13.7   Representing risk on Gaussian distribution of requirement volatility.

Without referring to the Gaussian tables, we can compute the tail area using Excel function in the following expression:

Right tail = 1 – NORMDIST (USL, mean, standard deviation, 1)

Substituting our values, we get right tail = 0.04697 or 4.697%.

You can measure opportunity with the same yardstick that measures the risk involved. They go together.

Earl Nightingale

The remaining area under the Gaussian measures the probability of meeting the goal or "capability." In our case, requirement volatility capability is 95.3%, as marked in Figure 13.7.

Capability and risk are complementary. If one is absent the other steps in.

As an extension of the risk calculation procedure, we can calculate risks for tails based on their distances from the mean. As an example, the tail areas are calculated for a few useful values of distance from mean and given in Table 13.1.

Table 13.1 contains the solution to the one-tailed problem and presents the probability of processes exceeding a given specification limit. Several one-tailed problems, such as the probability of defect density exceeding an upper limit, are the probability of productivity falling below a lower specification limit.

There are several two-tailed problems. These processes have both an upper specification limit and a lower specification limit. For the effort variance metric, the specification limits are ±20% in a certain enhancement project. The actual performance is characterized by a normal distribution with mean = 14 and standard deviation = 15. The two specification limits define two tails.

The Excel syntax for the previous computation is as follows:

Left tail = NORMDIST (LSL, mean, standard deviation, 1)
Right tail = 1-NORMDIST (USL, mean, standard deviation, 1)
Total risk in the process = left tail + right tail

The calculations are shown in Data 13.1.

The left tail involves process compliance risk. When teams save, there is a risk of adopting short cuts, which might later boomerang as product failure. The right tail has a plain cost risk. The total risk in the project could be the sum of the two-tailed areas. Sometimes, the two tails can attract different weights, for a "weighted" sum calculation of total risk. We have used a plain summation in Data 13.1 with the following result:

Table 13.1 Risk Calculation for Tails

Distance from Mean	Tail Area	% Tail Area
0.2	0.420740	42.074
0.4	0.344578	34.458
0.6	0.274253	27.425
0.8	0.211855	21.186
1.0	0.158655	15.866
1.2	0.115070	11.507
1.4	0.080757	8.076
1.6	0.054799	5.480
1.8	0.035930	3.593
2.0	0.022750	2.275
2.2	0.013903	1.390
2.4	0.008198	0.820
2.6	0.004661	0.466
2.8	0.002555	0.256
3.0	0.001350	0.135

Left tail = 0.0117
Right tail = 0.3446
Total risk = 0.3563

In previously mentioned risk analysis, cost escalation seems to be the dominating risk.

Data 13.1 Two-Sided Risk Estimation

Metric		Effort Variance (%)
Historic Data		
	Mean	14
	Sigma	15
Specification Limits		
	USL	20
	LSL	−20
Left tail		0.011705
Right tail		0.344578
Total risk		0.356284

BOX 13.3   GAUSSIAN SMOOTHENING

In image reconstruction Gaussian distribution is used.

In the domain of electromagnetic radiation, antenna beam widths are Gaussian reconstructed from the half power beam widths, which are easier to measure. The empirical construction of the beam with multiple data points is time consuming and looks less attractive when Gaussian smoothening is an accepted scientific practice. Gaussian smoothening saves time and money, and yet succeeds in constructing truth. In image processing, Gaussian smoothening is widely used.

An example of a common algorithm used to perform image smoothening is Gaussian. Each pixel is convolved with a Gaussian kernel and summed up; the result is suppression of noise, better signal-to-noise ratio, and better quality image. The bell curve is used to beat noise.

In digital signal processing, the Gaussian filter retrieves truer signals. In spatial smoothening MRI images, Gaussian smoothening is used to enrich the picture. Gaussian smoothening blurs the noise. The degree of smoothening is determined by the standard deviation of the Gaussian. Larger standard deviation Gaussians, of course, require larger convolution kernels to be accurately represented. "The Gaussian outputs a 'weighted average' of each pixel's neighborhood, with the average weighted more towards the value of the central pixels."

During the reconstruction of scanned images, Gaussian smoothening is like a low-pass filter. The Gaussian window is an attractive option for volume visualization in CT scans [8].

Combining Normal Probability Density Functions (PDF): The Law of Quadrature

A very useful property of the normal distribution is that we can easily combine several normal PDFs using a simple rule set:

Add the means to obtain the overall mean.
Add the variances to obtain the overall variance.

For example, the schedule performance of milestones can be combined using this property. The overall schedule for the project is the sum of schedules of milestones. The overall variance in the project schedule is the sum of individual milestone schedule variances. An example is available in Table 13.2.

The root cause for risk is variance, and Table 13.2 provides variance data across the project at every declared milestone. These milestones constitute on the critical path. Their variances are added by using the law of quadrature to obtain the overall

Table 13.2 Milestones Schedule Estimates

Milestone No.	Milestone	Estimated Schedule Days			
		Mean	Cumulative	Sigma	Variance
1	Start	0	0	3	9
2	Requirement gathering	3	3	1	1
3	Requirement documentation	4	7	1	1
4	High level design	12	19	2	4
5	Detailed design	20	39	2	4
6	Code for selected 10 modules	40	79	3	9
7	Code for next 10 modules	60	139	3	9
8	Code for remaining modules	30	169	2	4
9	System test	15	184	3	9
10	Integration test	12	196	2	4
11	User acceptance test	12	208	1	1
12	Finish	3	211	3	9
	Overall mean	211			
	Overall variance				64
	Overall sigma (sqrt of var)			8	

Note: The milestones above are on the critical path.

variance in the project. The perception of risks pervades the project during its life cycle. Figure 13.8a shows the bell curves for milestone deliveries.

The overall variance is filled in black. If the milestones were nonnormally distributed, the overall variance in a project is typically obtained by a procedure called *Monte Carlo simulation*. The overall variance is obtained in an elegant and simple method because we assume normal distribution in our case.

The overall project delivery variance and mean are shown separately in Figure 13.8b for further analysis. The delivery day is marked as 226, and the tail beyond is known as

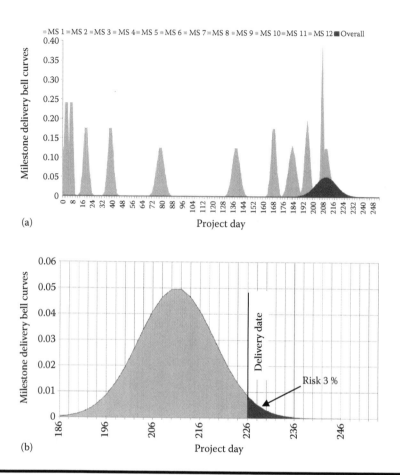

Figure 13.8 (a) Milestone delivery bell curves and (b) project delivery bell curve.

project schedule risk. From the Gaussian formula, risk can be computed as 3%. From the graph, we can also see the following three PERT values used in project estimation:

Optimistic value = 186
Pessimistic value = 236
Most likely value (mean) = 211

Using the golden rule, we can also estimate the delivery date, as follows:

$$\text{Delivery date} = \frac{t_{opt} + t_{pess} + kt_{most\ likely}}{6}$$

where k is a constant, normally taken as 4 but can be changed depending on the nature of the project.

We feel that the above example illustrates the most admirable capability of Gaussian distribution. It has made simulation a transparent job, which otherwise stays as a black box technique using sophisticated but less understood tools.

An Inverse Problem

Let us revisit Equation 13.2, which defines the Gaussian PDF. The integration of this PDF leads to the cumulative distribution function (CDF), $F(x)$, defined as follows:

$$F(x) = \frac{1}{\sqrt{2\pi}} \int_{-\infty}^{x} e^{-\left(\frac{x-\mu}{\sigma}\right)^2 / 2} \, dx \tag{13.4}$$

This CDF can be plotted using Excel NORMDIST(x, mean, SD, 1). As an example, we plot the CDF for requirement volatility with mean = 3.3% and standard deviation = 4% in Figure 13.9. The inverse problem is given that the Gaussian $F(x) = 0.78$; what is x? This question and its answer are shown in Figure 13.9. The answer is marked as 6.39. This solution has been obtained graphically.

We can use the Excel function NORMINV to find x, as follows:

NORMINV(0.78, 3.3, 4) = 6.3888

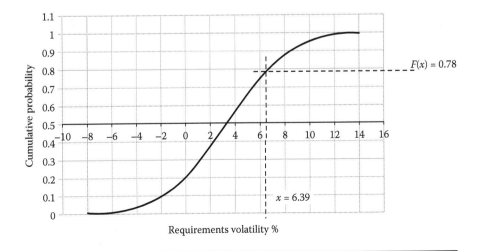

Figure 13.9   Gaussian cumulative distribution function of requirements volatility.

On the same metric, we can pose another inverse problem. If we reject 5% of events on either side of the Gaussian distribution, what is the range of requirement volatility within the acceptable region? The upper rejection point has a cumulative probability of 0.95 and the corresponding x value is as follows:

NORMINV(0.95, 3.3, 4) = 9.879%

The lower rejection point has a cumulative probability of 0.05, and the corresponding x value is as follows:

NORMINV(0.05, 3.3, 4) = –3.279%

Therefore, requirement volatility has a range defined by the interval [–3.279, 9.879] when 5% of events are rejected in both tails. Sometimes it helps to discard extreme ends of the mathematical function and consider the truncated interval as the practical dispersion. User's judgment is required to decide on how much to cut off. Typically, people choose any one of 1%, 5%, and 10% truncations. The truncated range accordingly shrinks in progression. When it comes to determining range, it depends on how the problem gets formulated. Whatever be the formulation, the Gaussian can provide a simple and ready answer.

The formulation of the problem is often more essential than its solution, which may be merely a matter of mathematical or experimental skill.

Albert Einstein

BOX 13.4   IS THERE AN AVERAGE MAN?

A landmark in the history of the bell curve is the notion of the average man put forward by Lambert Adolphe Jacques Quetelet (1796–1874), a Belgian astronomer, mathematician, statistician, and sociologist. He applied the bell curve to social science, which he called a *social physics*. He collected the heights of 100,000 French conscripts and the chest measurements of 5738 Scotch soldiers. The probable error in these measurements was approximately 2 inches. He found "harmonious" variations around the average. There was an astonishing symmetry and also an inevitable mixture. It looked as if there existed a fictitious average man or ideal man and others mere deviations from the ideal. Variations came from "constant causes" and some extreme perturbations came from "accidental causes." (We are reminded of Shewhart's common causes and special causes, a profound idea that would appear in 1920, a hundred or so years later, in his statistical process control.)

Quetelet was convinced that "there is a general law which dominates our universe." He presented a most important and extensive role for the average man. The physician could thus determine the most useful remedies and the action to be taken, in both usual and unusual cases, by comparing with the fictitious average man. Hence, the artist could predict truth, the politician could predict public sentiments, the naturalist could predict racial types, and social scientists could predict laws of birth, growth, and decay.

Quetelet compared the average man with the center of gravity. Everything is to be viewed as varying about a normal state in a manner to be accurately described by beautiful bell-shaped curves of perfect symmetry but of varying amplitude. Thus, it is that the individual varies about his normal self and the members of a group vary about their average. In social physics, the bell curve represents the true mechanics of human history.

The average man is free from excess and defect. Nature is striving to produce the average man but fails because of the interference from a multitude of causes [9].

Lesson learned: By analogy, likewise, the industry strives to achieve ideal processes but fails because of interferences.

Growth should be judged by averages; variations must be used to detect problems.

Process Capability Indices

A process is said to be capable if two conditions are met: it should show less variation, and it should be aligned to goal.

Variation must be contained inside the specification window available. The practical range of the process (truncated) must be less than the gap (USL – LSL). How much of the specification window is consumed by the process determines capability. Hence, we use the following two equations in assessing the first index, known as C_p, process potential:

Specification window = USL – LSL

C_p, process potential = specification window/6σ

Alignment depends on how close the process mean is to the process target. Alignment can be measured by the distance between process mean and process target. Processes tend to drift away from the target. Drift diminishes capability. A drifted process achieves only a part of its potential. Hence, we have the concept of achieve process capability index C_{pk}, shown as follows:

$$C_{pk}, \text{ achieved process capability} = C_p(1 - k)$$

where k = process drift/half of specification window.

The more the drift, the more will be the value of k and the less will be the value of C_{pk}.

There is an established tradition that puts C_p at 1.0 for acceptable quality and 2.0 for excellent quality.

Besides, if $C_{pk} < C_p$, then the process requires alignment to the target.

An example calculation is shown in Table 13.3. Effort variance metric is used for this example calculation. The mean is 5%, and the SD is 7%. Specification limits typically are ±10%. In Table 13.3, C_p is calculated as 0.48, and C_{pk} is calculated as 0.24. Two improvement opportunities emerge from these calculations. The first opportunity is to improve C_{pk} and make it equal to C_p; this involves process alignment, meaning a shift in the central tendency. The next opportunity is to achieve

Table 13.3 Process Capability Indices

Process Metric	Effort Variance	%
Process Goals		
Target	0	%
USL	10	%
LSL	−10	%
Process Performance		
Mean	5	%
Sigma	7	%
Process Capability Indices		
C_p	= (USL − LSL)/6σ	0.48
k	= (Drift)/(0.5*(USL − LSL))	0.50
C_{pk}	= C_p*(1 − k)	0.24

a breakthrough reduction in process variation and hence increase C_p. Reducing variation is more challenging than shifting process mean.

Reduce variation. Knowledge about variation is profound knowledge.

Deming

z Score Calculation

Process z is defined in Equation 13.5. The difference between observed value and ideal mean is divided by standard deviation to obtain process z:

$$z = \frac{x - \mu}{\sigma} \tag{13.5}$$

The previously mentioned formula is also known as z score. Larger deviations from mean earn larger values of score. Hence, z score is a metric of deviation from mean. Because z score is normalized, it is dimensionless.

In practice, what is measured is deviation from the target. Hence, practical z score has the following formula:

$$z = \frac{x - T}{\sigma} \tag{13.6}$$

For each project metric, z score can be computed as shown in the example in Table 13.4. Six development project metrics are considered in the table, and using Equation 13.6, the z scores have been computed.

Table 13.4   *z* Score

Metric	Target	Sigma	Performance	z Score
Effort variance	0	5	10	2.00
Schedule variance	0	3	5	1.67
Scope creep	0	2	3	1.50
Defect density	0	1	3	3.00
Complexity	30	10	70	4.00
CSAT	8	2	5	−1.50

The choice of metrics strikes a balance between process, product, and business objectives.

Process	Effort variance
	Schedule variance
	Scope creep
Product	Defect density
	Complexity
Business	CSAT (customer satisfaction)

In this example, the larger the score, the greater the deviation from the target. The bar length indicates the statistical distance of mean from targets, the magnitude of problems. The advantage is that all metrics performance can be shown in the same chart with a common unit (Figure 13.10).

The picture provides a balanced view of the development project.

The approach of measuring statistical distances using z scores can be used to compare current year performance from last year performance. Such a comparison is shown in Table 13.5. The tornedo chart is shown in Figure 13.11.

Negative z scores indicate the statistically significant reduction in problems. The positive z score of CSAT spells significant improvement. The usual practical

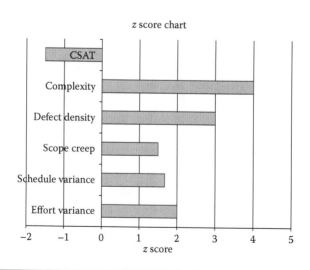

Figure 13.10 *z* Scores: statistical distance from targets.

Table 13.5 Improvement Scores

Metric	Organization Sigma	Last Year Mean	Current Year Mean	Shift in Mean	z Score
Effort variance	5	12	10	–2	–0.40
Schedule variance	3	6	5	–1	–0.33
Scope creep	2	6	3	–3	–1.50
Defect density	1	3	2	–1	–1.00
Complexity	10	70	60	–10	–1.00
CSAT	2	5	6	1	0.50

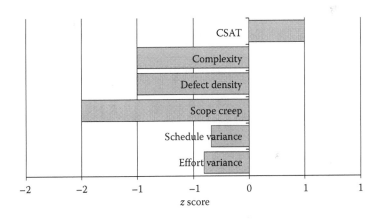

Figure 13.11 Score improvement.

question asked, when improvements are reported, is whether the improvements are statistically significant. The direction of z scores indicate improvement and the magnitude indicate statistical significance.

Sigma Level: Safety Margin

As the story goes, in 1986 Bill Smith, the father of Six Sigma, showed a bell curve to the CEO Bob Galvin and proved that there is a finite probability that processes cross the limits and explained why defects reach customers despite multiple testing. Six Sigma began as a reliability model. Jack Welch made Six Sigma a discipline in GE and later opined, "the mean is fine it is the standard deviation that spells trouble."

The entire Six Sigma methodology depends on the normal distribution. Assuming normal distribution, we can calculate sigma level of safety margin to the customer using the following definition:

$$\text{Sigma level} = (\text{USL} - \text{mean})/\sigma, \text{ or}$$

$$= (\text{Mean} - \text{lSL})/\sigma$$

$$\text{Whichever is smaller} \qquad (13.7)$$

The sigma level can be converted into risk using ideas developed early in this chapter. Risk is expressed in parts per million or defects per million opportunities (DPMO). For example, we can consider two-tailed problems and find tail areas for different distances of tails from the mean: 1, 2, 3, 4, 5, and 6 sigmas as shown in Table 13.6. The distances are known as sigma levels. The tail areas are known as defect levels. In Table 13.6, defect levels are presented first in fractions, then in percentage, and finally in PPM or DPMO in the last column.

Table 13.6 Six Sigma Conversion Table

Sigma Level	Tail Areas		
	Fraction	%	PPM (DPMO)
Part A: Pure Scale			
1	0.3173105078629	31.731050786	317,310.50786
2	0.0455002638964	4.550026390	45,500.26390
3	0.0026997960633	0.269979606	2699.79606
4	0.0000633424837	0.006334248	63.34248
5	0.0000005733031	0.000057330	0.57330
6	0.0000000019732	0.000000197	0.00197
Part B: Practical Scale (1.5 Sigma Drift Included)			
1	0.6976721	69.76721	697,672.13
2	0.3087702	30.87702	308,770.17
3	0.0668106	6.68106	66,810.60
4	0.0062097	0.62097	6209.68
5	0.0002326	0.02326	232.63
6	0.0000034	0.00034	3.40

Table 13.6 has two parts. In part A, the tail areas have been computed the usual way. This suggests a defect level of 1.97 parts per billion for a process to qualify as a Six Sigma process. This requirement was contested internally in Motorola by pragmatic managers who wanted an allowance for long-term drifts in processes. In particular, they wanted the "Shewhart allowance of 1.5 sigma drift in control charts" to be made available in Six Sigma considerations. Part B of Table 13.6 has been computed with Shewhart allowance; this table suggests 3.4 PPM defects for a Six Sigma process, approximately 1726 times more defects than the pure scale. Finally, the practical scale has prevailed and is widely used as a quality standard.

In Six Sigma culture, we respond to mathematically derived tail probabilities even if there is no physical event in the tail region.

Statistical Tests

Gaussian properties are extensively used in statistical tests to compare results. If two processes are represented by adjacently located Gaussian curves and if the tails do not overlap, they are distinctly different processes. If tails overlap, perhaps they are not so different. To resolve this problem, we resort to statistical tests, such as z test, t test, and F test. In all tests, we find a p value, the probability of finding one sample from the other lot. To calculate p value, in commonly used statistical tests, the Gaussian curve is used.

We have seen how the bell curve can be put to a variety of applications in software engineering and management.

**BOX 13.5   ELECTRON CHARGE TO
MASS RATIO MEASUREMENT**

Nobel laureate J. J. Thomson measured the invisibly small particle electron in 1897 at the well-known Cavendish laboratory in Cambridge, England.

By carefully measuring how the cathode rays were deflected by electric and magnetic fields, Thomson was able to determine the ratio between the electric charge (e) and the mass (m) of the rays. Thomson's result was

$$e/m = 1.8 \times 10^{-11} \text{ coulombs/kg}$$

He received the Nobel Prize in 1906 for the discovery of the electron, the first elementary particle.

Nobelprize.org

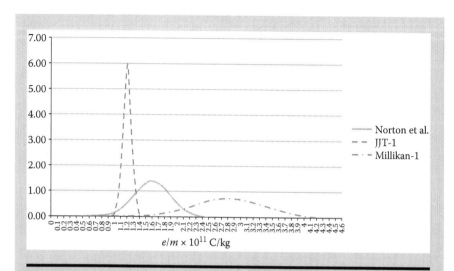

Figure 13.12 Gaussian error curves.

In reality, measurements do not get reported by a single value. We have either a range of values or a mean with associated standard deviation. We are discussing *e/m* measurements of electrons in coulombs per kilogram. The numbers presented here must be multiplied by 10^{11}.

Todays' accepted mean *e/m* is 1.758820088 with a standard deviation of 0.000000013.

Earlier trials by J. J. Thomson gave results between 1.1 and 1.4 [10]. Norton et al. [11] have shown results with a mean value of 1.60 and a standard deviation value of 0.29. Earlier attempts by Millikan [12] result in a mean value of 2.82 and a standard deviation of 0.55. These three results are shown as reconstructed Gaussian curves in Figure 13.9.

The three bell curves indicate measurement reliabilities available in those experiments. The broader the curve, the less the measurement reliability. It may also be noted that broader curves have shorter peaks. In this context, one can intuitively feel that the height of the peak can also be considered as an indicator of measurement reliability. Narrower curves indicate "precision" in measurement (Figure 13.12).

Review Questions

1. What are the various names by which the normal distribution is known?
2. What is Gaussian smoothening?

3. Why does the bell curve perfectly fit all the different variance metrics such as effort variance, schedule variance, and size variance?
4. Name the famous mathematicians and scientists who have contributed to the development of the bell curve.
5. Mention three important applications of the bell curve.

Exercises

1. In the context of a bell curve with mean = 7 and standard deviation = 6, find the z score of a data point 25.
2. Dr. Shewhart prescribed three sigma limits to control charts. What are the tail areas outside these limits?
3. Find the area under the bell curve included inside two sigma limits.
4. Find the percentage of area beyond six sigma limits. Express this in parts per million.
5. A process peaks at 4. Its specification limits are 2 and 5. If the standard deviation is 1, find the process capability indices C_p and C_{pk}.

References

1. A. de Moivre, *The Doctrine of Chances: Or, a Method of Calculating the Probability of Events in Play*, W. Pearson, London, 1718.
2. J. C. Friedrich Gauss. Available at http://en.wikipedia.org/wiki/Carl_Friedrich_Gauss.
3. H. Fischer, *A History of the Central Limit Theorem: From Classical to Modern Probability Theory*, Springer, New York, Dordrecht, Heidelberg, London, 2010.
4. Sir F. Galton, *Natural Inheritance*, MacMillan, 1889.
5. G. Pólya, Über den zentralenGrenzwertsatz der Wahrscheinlichkeitsrechnung und das Momentenproblem, *MathematischeZeitschrift* (in German), 8(3–4), 171–181, 1920.
6. L. Le Cam, The central limit theorem around 1935, *Statistical Science*, 1(1), 78–91, 1986.
7. D. J. Poirier, *Intermediate Statistics and Econometrics: A Comparative Approach*, MIT Press, London, Cambridge, 1995.
8. Elvins, *Volume Visualization*. Available at http://www.cs.rug.nl/~michael/FANTOM/FANTOM1a.pdf.
9. F. H. Hankins, *Adolphe Quetelet as Statistician*, Columbia University, New York, 1908.
10. P. F. Dahl, *Flash of the Cathode Rays: A History of JJ Thomson's Electron*, CRC Press, 1997.
11. M. Norton, C. Bush, B. Atinaja and B. Steven, *Electron Charge to Mass Ratio*. Available at http://www.siue.edu/~mnorton/Ratio.pdf.
12. R. A. Millikan, *The Electron: Its Isolation and Measurements and the Determination of Some of Its Properties*, The University of Chicago Press, Chicago, 1917.

Suggested Readings

Adams, W. J., *The Life and Times of the Central Limit Theorem*, American Mathematical Soc., 2nd edition, 2009.

Hauser, H., E. Groller and T. Theussl, *Mastering Windows: Improving Reconstruction*, Institute of Computer Graphics Vienna, University of Technology. Available at http://www.cg.tuwien.ac.at/research/vis/vismed/Windows/MasteringWindows.pdf.

Jones, R., *What's Who?: A Dictionary of Things Named after People and the People They Are Named After*, Troubador Publishing Ltd., UK, 2009.

Leiter, D. J. and S. Leiter, *A to Z of Physicists*, Infobase Publishing, 2009.

Millikan, R. A., Wikipedia, the free encyclopedia. Available at http://en.wikipedia.org/wiki/Robert_Andrews_Millikan.

R'Udt, D., *The Central Limit Theorem*, University of Toronto, 2010. Available at http://wiki.math.toronto.edu/TorontoMathWiki/images/0/00/MAT1000DanielRuedt.pdf.

Tijms, H., *Understanding Probability: Chance Rules in Everyday Life*, Cambridge University Press, p. 169, ISBN 0-521-54036-4, UK, 2004.

Watkins, T., *The History of the Central Limit Theorem*, San José State University, Silicon Valley & Tornado Alley. Available http://www.applet-magic.com/randovar.htm.

Chapter 14

Law of Compliance: Uniform Distribution

In sharp contrast to the bell curve, the uniform distribution looks plain, flat, and simple but has some interesting applications. There are no tails in the uniform distribution. There is no peak either. Uniformly distributed processes are rare in manufacturing, but nearly uniformly distributed processes exist in IT services.

Uniform distribution is a continuous distribution bounded between two limits, A and B. The probability density function (PDF) may be stated as follows:

$$F(x) = \frac{1}{A - B} \quad \text{for } A \leq x \leq B \tag{14.1}$$

The plot of uniform distribution, shown in Figure 14.1, is a rectangle.

Hence, uniform distribution is also known as the rectangular distribution. It may be recalled that by integrating a PDF, we get a cumulative distribution function (CDF). It is also known that when we integrate a rectangle, we get a triangle. In this case, the PDF is rectangle. Integrating the PDF, we get a triangle which is the CDF. Obviously, the CDF is a triangle.

The PDF has the following statistics:

Mean = $(A + B)/2$
Median = $(A + B)/2$
Range = $B - A$
Variance = $\dfrac{(B - A)^2}{12}$

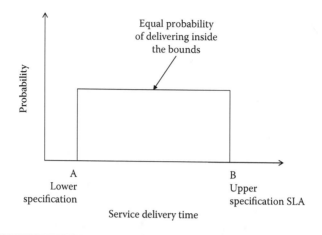

Figure 14.1   Probability density function (PDF) of uniform distribution.

$$\text{Standard deviation} = \sqrt{\frac{(B-A)^2}{12}}$$

Skewness = 0
Kurtosis = 9/5
Relative kurtosis = −1.2

The CDF is defined as follows:

$$F(x) = \frac{x-A}{B-A} \qquad \text{for } A \le x \le B \tag{14.2}$$

The CDF is a triangle, as shown in Figure 14.2.

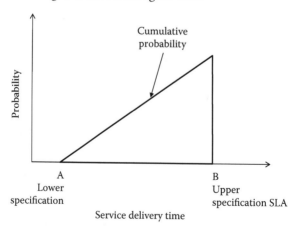

Figure 14.2   Cumulative distribution function (CDF) of uniform distribution.

Bounded Distribution

A principal feature of the uniform distribution is its boundary. It is bounded on both sides, in sharp contrast with the unbounded bell curve. To qualify as a uniform distribution function, three criteria must be met:

1. Bounding limits must be "final minimum," clear, and not elastic or vague
2. Containment criterion 100%
3. Uniform probability criterion

In practice, there could be challenges in meeting criterion 3; there could be turbulence between the limits.

There are many bounded phenomena in software projects. Customer satisfaction data are bounded between 1 and 5. SLA compliance is bounded between 0% and 100%. Time to repair is bounded between 0 and agreed upon maximum time; hence, criterion 1 is met. The question in criterion 3 is uniformity. The process could be any other bounded distributions, beta or truncated Gaussians. Here comes a need for approximation and simplification. Uniform distribution is simpler and hence is preferred in most cases.

When the customer does not specify boundary, and it is left to the process QA to define the boundary based on data, challenges arise. In particular, estimating the upper bound could be a challenging problem. Data may not show a sharp edge. If we are using only sample data, then the problem of fixing upper bound is analogous to the German tank problem relevant to the World War II situation [1].

German tanks were produced according to a uniform distribution [1, N]. The number of tanks captured was n, a mere sample. Can we estimate N from the sample data?

$$N = m - 1 + \frac{m}{n} \qquad (14.3)$$

where N is the upper bound of the uniform distribution [1, N], n is the number of tanks captured, and m is the Largest serial number in the sample.

Later, after the war, statistical estimations were found to be much closer to the truth than intelligence reports, four to one.

Random Number Generators

Random numbers follow uniform distribution. However, it is difficult to generate random numbers that fit perfectly into a uniform distribution. There are several random number generators (RNGs), but they predict uniform distribution with varying levels of success. For reliable results in simulation, we need perfect

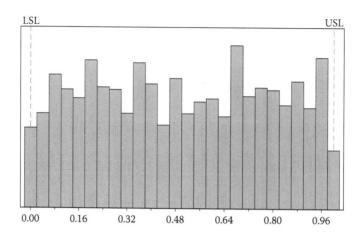

Figure 14.3 Distribution of random numbers.

random numbers, which means their PDF must be super uniform. A few RNGs aim at smooth uniform PDFs, whereas others accommodate deviations from perfect smoothness. To illustrate this point, Figure 14.3 shows a histogram of 1000 random numbers generated by Excel function RAND(). We can see a histogram with perturbed profile, bounded between 0 and 1. These are pseudorandom numbers, as best as a practical tool can predict.

Shuttle Time

If a shuttle bus has a cycle time of 40 minutes, the waiting time is uniformly distributed between 0 and 40. A passenger who reaches the bus stand in a random hour and is unaware of when the last shuttle left is bound to wait up to 40 minutes. From his viewpoint, the arrival of the shuttle is uniformly distributed from 0 to 50 minutes.

Parkinson's Law

If a service manager assigns a task to his team and suggests that the team completes the task within an interval of 4 to 6 days, the team will take anywhere between 4 and 6 days. The team will exploit the goal window and not exercise its own natural capability to finish the task in a naturally possible time. For all we know, the team may be able to finish the job in less than 4 days, but human behavior is to stretch the job according to Parkinson's law which states,

Work expands to fill the time available.

If teams exercised their natural capabilities, the PDF of completion time would have a peak. Because the team negotiates time to meet specified goals, we end up with a performance that does not have character.

Censored Process

If components out of specifications are removed, the reaming lot shows nearly uniform distribution between the specification limits. The censored lot tends to be more uniform if the original lot shows wide variation and if the specification limits are stringent. An example is when a semiconductor component manufacturer screens best pieces from the line and sells them at premium prices as close tolerance devices. He downgrades the rejected components and sells them at a lower price. The premium components after censoring show uniform distribution.

Perfect Departure

From an auditorium, if people leave in perfect queues, the departure is uniformly distributed. The probability of people crossing the gate is uniformly distributed between 0 and a finite time that depends on the number of people and width of the gate. When people try to break the queue and rush out, the departure is skewed, the worst case being a stampede.

Estimating Calibration Uncertainty with Minimal Information

During the calibration of measuring equipment, we need to assess uncertainties. There are two types of uncertainties affecting measurement: type A is determined from data, and type B is guessed. In type B estimates, we might only be able to estimate the upper and the lower limits of uncertainty. We have to assume the value is equally likely to fall anywhere in between, that is, a rectangular or uniform distribution. The standard uncertainty for a rectangular distribution is equal to $\frac{a}{\sqrt{3}}$, where a is the semirange between the upper and the lower limits.

Estimation of uncertainty using the uniform distribution is relevant in the following cases:

- Digital resolution uncertainty
- RF phase angle
- Quantization error
- As an expression of ignorance

From the above examples, it can be seen, inspite of simplicity, the uniform distribution can serve as a handy model to represent several real time processes.

BOX 14.1 AIRPORT TAXI-OUT TIME

Flight delay has been and continues to be one of the most critical problems for airports. A large percentage of flight delays occur on the ground. Among all the delays, historical data indicate that taxi-out times contribute to more than 60% of the total. The taxi-out time is defined as the ground transit time between the pushback time scheduled or updated by airlines and the takeoff time when the aircraft is captured by the radar tracking system. A queuing model was introduced to estimate the taxi-out time at Logan Airport. The takeoff queue size was defined as the number of takeoffs that take place between the aircraft pushback time and its takeoff time.

A histogram of taxi-out time is flat, suggesting uniform distribution. The cumulative frequencies form a straight line, confirming the assumption of uniform distribution. The straight line regresses, with an R^2 value of 0.98.

Extreme taxi-out times occur due to bad weather. These are not included in the construction of uniform PDF. Other models have an average prediction error of three minutes, whereas uniform distribution PDF prediction error is less than 1 minute [2].

The PDF is reconstructed in Figure 14.4.

By analogy, the same uniform distribution is relevant to software support services. The response time in complex situations follows uniform distribution.

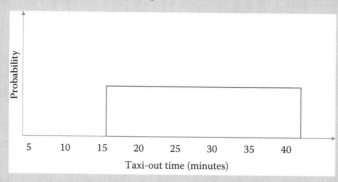

Figure 14.4 PDF of taxi-out time.

? Review Questions

1. What are pseudorandom numbers? How are they related to the uniform distribution?
2. What is the shape of the CDF of uniform distribution?
3. What is the uncertainty in a digital measuring device if its resolution is ±1 mV? We assume uniform distribution here.
4. What is the famous German tank problem?
5. What is the formula for variance in uniform distribution?

Exercises

1. Calculate kurtosis in uniform distribution if $A = 1$ and $B = 2$.
2. Solve the German tank problem if 30 tanks were captured and the largest serial number is 115. That means you have to estimate the number of tanks produced in Germany. Clue: use Equation 14.3.
3. Calculate the median of the uniform distribution with $A = 2$ and $B = 3$.

References

1. D. Evans, *The German Tank Problem*, Rose-Hulman Institute of Technology, Terre Haute, Indiana. Available at https://www.causeweb.org/webinar/activity/2009-09/2009-09.pdf.
2. T. V. Truong, The distribution function of airport taxi-out times and selected applications, *Journal of the Transportation Research Forum*, 50(2), 33–44, Summer 2011.

Suggested Readings

Adams, T. M., *Guide for Estimation of Measurement Uncertainty in Testing*, A2LA Guide for Estimation of Measurement Uncertainty in Testing, July 2002. https://www.a2la.org/guidance/est_mu_testing.pdf.

Bell, S., *A Beginner's Guide to Uncertainty of Measurement*, Centre for Basic, Thermal and Length Metrology, National Physical Laboratory, Teddington, Middlesex, UK, August 1999. https://www.wmo.int/pages/prog/gcos/documents/gruanmanuals/UK_NPL/mgpg11.pdf.

Castrup, H., *A Critique of the Uniform Distribution*, Integrated Sciences Group, 2000.

Hellekalek, P., Good random number generators are (not so) easy to find, *Mathematics and Computers in Simulation*, 46(5–6), 485–505, 1998.

Larsen, R. J. and M. L. Marx, *An Introduction to Mathematical Statistics and Its Applications*, Prentice Hall, Upper Saddle River, NJ, 2006.

Chapter 15

Law for Estimation: Triangular Distribution

Triangular distribution is one of the simplest known distributions. It can be constructed even with minimum or subjective data. The triangular distribution is marked by its sharp limits. Inside the area defined by the limits, the triangle offers a peak that can assume any position with the limits.

These limits are Limits of Probability. Once the limits are known, filling the distribution is easy.

> The scientific imagination always restrains itself within the limits of probability.
>
> **Thomas Huxley**

It is also an approximate distribution: The comfort of approximation is matched by convenience of usage and freedom. This freedom offered by the triangular distribution makes it a favorite choice during business decision making and simulation. Approximation is the hallmark of genius.

> It is the mark of an educated mind to rest satisfied with the degree of precision which the nature of the subject admits and not to seek exactness where only an approximation is possible.
>
> **Aristotle**

Three points that define a triangle have the power to guide. They seem to stem from a cosmic design of reality (see Box 15.1).

The earliest record of triangular distributions seems to be in 1755 (about one century after the discovery of the related and slightly more sophisticated beta distribution). Recently, the triangular distribution has been used as a proxy for the beta distribution. However, the triangular distribution is simpler and more effective [1].

Bell Curve Morphs into a Triangle

We can consider the normal distribution to have a principal body with tail adjuncts. The truncation points divide these two ingredients, as shown in Figure 15.1. The triangle is a good proxy to the curved but truncated principal body.

The above illustration presents the advantages of the triangular proxy; it has focus and simplicity. The tails are discarded, and curvature is replaced by linear construction. In project management, the tails represent risk and the body relates to primary delivery. In risk management, we pursue distribution with tails having

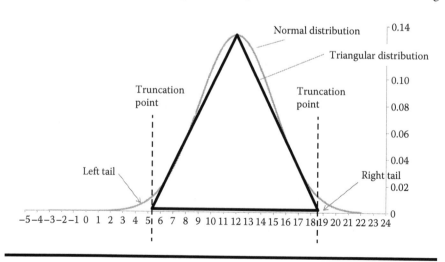

Figure 15.1 Normal distribution as triangle.

appropriate characteristics to suit the given problem, such as the exponential and Pareto for long tails and Gumbel and Weibull for constrained tails. In delivery management, the triangle stands for the core output, with risks being relegated to the next layer of abstraction and kept invisible, for the present, to the manager.

BOX 15.1 THREE POINTS FOR GUIDANCE

It is good to know that as a general navigation concept, guidance is available from three points of reference. If you have three, you get your bearing. Here is a related quote:

> Sailors and seafarers find their bearings at sea by means of natural points of reference located along the coast. These points, for example church spires, hills, water-towers or lighthouses that generally stand out from the rest of the coastline, are called amers [seamarks or landmarks]. All you have to do is identify three such landmarks in complementary directions so as to be able to construct a triangle which inevitably contains your ship. This triangle drawn on the navigation map is called the "triangle of uncertainty."
>
> **Cécile Le Prado**

Likewise, in software development, project three-point estimates provide great guidance. Managers can navigate through the project life cycle with this help.

Mental Model for Estimation

The triangular distribution is both bounded and peaked. Hence, it carries the advantages of being bounded like the uniform distribution and has the additional advantage of having a strong central tendency. The triangular distribution is a simplified model of process results and a pragmatic substitute for normal and other distributions. The project management variables of time, cost, and performance can be elegantly modeled using the triangular distribution.

A plot of the triangle may be seen in Figure 15.2. The bottom edges *a* and *b* represent optimistic and pessimistic values. The middle edge *c* represents the most likely value. This is the estimation triangle extensively used in management.

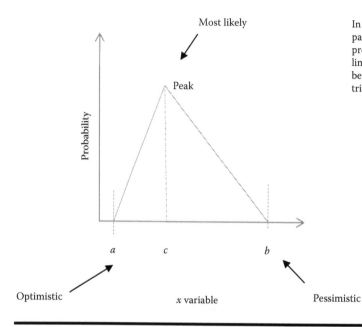

Figure 15.2 Triangular estimation—a widely used mental model.

The probability density function (PDF) is split between simple equations to the two straight lines to the peak from the two base points. The equation to the first line, from the left base point to the peak, is given as follows:

$$y(x) = \frac{2(x-a)}{(b-a)(c-a)}$$ (15.1)

The equation to the second line, from the peak to the second base point, is given as follows:

$$y(x) = \frac{2(b-x)}{(b-a)(b-c)}$$ (15.2)

Mean

The model statistics are derived from elementary geometrical properties of a triangle. Arithmetic mean is according to the following equation:

$$\text{Mean} = \frac{a+b+c}{3}$$ (15.3)

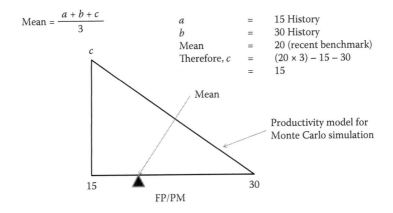

Figure 15.3 is represented with:

$$\text{Mean} = \frac{a + b + c}{3}$$

a	=	15 History
b	=	30 History
Mean	=	20 (recent benchmark)
Therefore, c	=	$(20 \times 3) - 15 - 30$
	=	15

Figure 15.3   Productivity model.

Here is an example of the equation application. Let us take the case of a project manager trying to give input to a simulator. Let us say he is required to define a PDF for productivity. From his own experience, he recalls that function point (FP)/person-month had varied from 15 to 30. He has also obtained a recent benchmark report claiming that the mean value of productivity is 20 FP/person-month. Following the Bayesian spirit, we can combine both these data, first a historic probability and then a recent piece of evidence. Substituting these three values in Equation 15.3, we derive the following:

a = 15
b = 30
Mean = 20
Therefore, $c = (20 \times 3) - 15 - 30 = 15$

This is a right-angled triangle skewed to the right, as shown in Figure 15.3. This model can now be used in Monte Carlo simulation as an input.

Median

If we are lucky enough to obtain the median value for a skewed result, then we can apply the formulas in Equations 15.4 and 15.5, as follows:

$$\text{Median} = a + \frac{\sqrt{(b - a)(c - a)}}{\sqrt{2}} \quad \text{when } c \geq (a + b)/2 \qquad (15.4)$$

$$= b - \frac{\sqrt{(b-a)(b-c)}}{\sqrt{2}} \quad \text{when } c \le (a+b)/2 \qquad (15.5)$$

The calculation is illustrated below by solving a problem.

QUESTION

Given the following inputs,

Minimum % SLA compliance = 50
Maximum % SLA compliance = 100
The median is 80%

Build a triangular PDF for service-level agreement (SLA) compliance and find where the peak occurs.

ANSWER

Let us rearrange the input information to suit our formulas.

$a = 50$
$b = 100$
Therefore, $(a + b)/2 = 75$
Median = 80

Substituting all these in Equation 15.5, we obtain

$$c = 84$$

Thus, all the three corners of the triangle are known. The model is plotted in Figure 15.4.

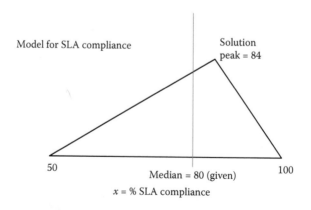

Figure 15.4   Triangular model of SLA compliance.

The triangular SLA compliance model is far superior to a plausible Gaussian model. Typically, the Gaussian tail would exceed 100%, distort calculations, and force us to take countermeasures such as a messy truncation. The triangular PDF is compact and does not outstep empirical experience.

Other Statistics

The mode is obviously the peak, and hence, the following relationship is true:

$$\text{Mode} = c$$

Dispersion is strongly indicated by the base width, $b - a$. However, a proper calculation of variance is according to the following equation:

$$\text{Variance} = (a^2 + b^2 + c^2 - ab - ac - bc)/18 \qquad (15.6)$$

An example is as follows:
Given

$a = 0,$
$b = 10,$
$c = 5$ (for a symmetrical triangular model),

We get

Variance = 4.17,
Standard deviation = 2.0412.

It may be noted that as c changes, the variance slightly changes.

Skew

Although the process boundaries constitute a firm base, the apex c can be moved from the left extreme to the right extreme, as shown in the three examples in Figure 15.5.

The first example has its peak at the lower limit and gives a triangle skewed to the right. The second example has its peak in the middle position between the limits, providing symmetry. The peak in the third example coincides with the upper limit, giving a negative skew. These three peaks demonstrate how the triangular PDF can be made to be symmetrical or skewed. The peak can take an infinite number of positions within these extremes.

c   Process mode
a   Lower boundary of process
b   Upper boundary of process

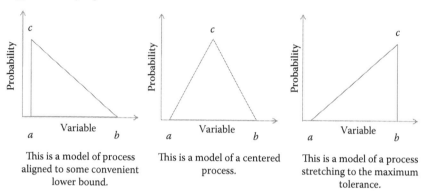

This is a model of process aligned to some convenient lower bound.

This is a model of a centered process.

This is a model of a process stretching to the maximum tolerance.

Figure 15.5   Triangular distribution—three examples.

The formula for skew is as follows:

$$\text{Skew} = \frac{\sqrt{2}(a+b-2c)(2a-b-c)(a-2b+c)}{5(a^2+b^2+c^2-ab-ac-bc)^{\frac{3}{2}}} \tag{15.7}$$

Equation 15.7 is used to construct a relationship between skew and mode c for a given $a = 0$ and $b = 10$. A graph of the relationship is plotted in Figure 15.6.

We can generate a wide range of skews using the relationship. In software development projects, the challenge arises in the form of skew. In a Gaussian-dominated statistical thinking, skew does not even exist. The triangular model provides a simple model to represent skew. Hence, the inherent advantages of the triangular model are threefold:

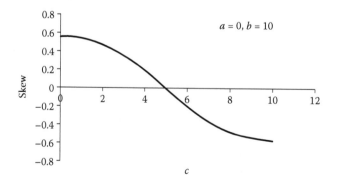

Figure 15.6   Relationship between Skew and mode c.

It shows a prominent central tendency.
It is capable of representing symmetry.
It is capable of showing skew, both left and right.

All these representations can be achieved with great agility.

Three-Point Schedule Estimation

Let us consider an example of schedule estimation by expert judgment for a software component development:

Optimistic value = 25 days
Pessimistic value = 50 days
Most likely value = 30 days

Applying the triangular PDF, the expected value of schedule is as follows:

$$\text{Mean} = (25 + 50 + 30)/3$$

$$= 35 \text{ days}$$

This may be compared with the conventional estimation technique using the program evaluation review technique (PERT) formula. The PERT formula will place the estimate as follows:

$$\text{PERT} = (t_o + 4t_m + t_p)/6$$

$$= (25 + 4 \times 30 + 50)/6$$

$$= 32.5 \text{ days}$$

It is seen that the triangular PDF gives a safer and more conservative estimate.

Beta Option

There have been interests in generalized triangles with curvature added. Wahed published "The Family of Curvi-Triangular Distributions" [2]. Brizz [3] has constructed two-faced triangles with one face a straight line and the second face exponential. However, the classic beta distribution provides smoothly curved bounded functions, and in the opinion of the authors, the good old beta distribution must be exploited first before experimenting with curvilinear versions of the triangle. A typical beta distribution model is shown in Figure 15.7. The problem of productivity is revisited with beta.

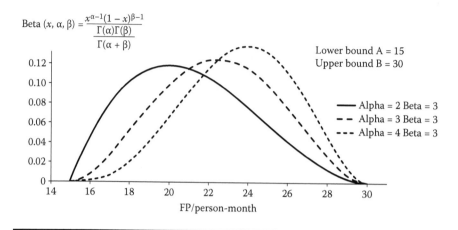

Figure 15.7 Beta distribution of productivity (FP/person-month).

For fixed upper and lower bounds 30 and 15, three curves have been drawn for three sets of shape parameters. The beta distribution bounded between 0 and 1 is defined as follows:

$$\text{Beta}(x, \alpha, \beta) = \frac{x^{\alpha-1}(1-x)^{\beta-1}}{\dfrac{\Gamma(\alpha)\Gamma(\beta)}{\Gamma(\alpha+\beta)}} \tag{15.8}$$

where Γ represents the gamma function, and α and β are the shape parameters.

Figure 15.7 looks more appealing than triangles, but the equation intimidates users. Beta distribution can offer an impressive array of bounded shapes. However, while implementing bounded functions in real-life projects, beta distribution was less acceptable despite its inherent power, and the "intuitive" triangular model was considered.

Triangular Risk Estimation

Like with any PDF, the triangle can be used for risk analysis. Figure 15.8 shows risk measurement in the triangular way.

Risk computation based on triangular function should be taken with far more seriousness and treated more urgently than risk measured based on tailed distributions. Tails in bell curves and other tailed distributions are mere extrapolations into extremes, whereas the triangle means business. Risk measured by typical triangular models should be specially treated because we do not generally anticipate risks in the triangular side of the world. Measuring risk with the body of a probabilistic distribution is very different from measuring risk with tails.

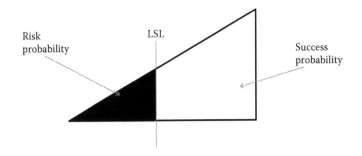

Figure 15.8   Risk estimation using triangular distribution.

Parameter Extraction

Precise parameters make precise models. In the case of the triangular model, we use expert judgment, Delphi, or wideband Delphi in the absence of data. Group judgment is more robust, resulting in a less skewed triangle. If data were available, we use histograms and visually match a triangle first and then, if called for, a least square error method after optimally binning the histogram. Going for MLE or other rigorous techniques is not warranted for triangular models. If more precise judgment of risk is involved, then the beta distribution should be chosen or the triangular model must be refined as van Dorp and Kotz [4] have done: They extended the model into a four-parameter version that allows even *J*-shaped forms. It may be seen that the TD offers all the facilities available in classical distribution models; it supports risk estimation and simulation; It is used in 3 point estimation and first order approximation of processes. All this is accomplished using elementary linear consideration.

BOX 15.2   A CRYSTAL CLEAR WORLD

The triangle offers a tailless view of process. This is a crystal clear world without the ambiguity. The presence of tails makes comparing two processes a complex affair; one often needs hypothesis testing and abstruse rules. It is quite plain with triangles, and one can make commonsense-based judgments.

For example, let us consider a productivity model defined by a triangle (30, 40, and 60) lines of code (LOC) per person day. To answer a question whether a productivity data point 61 belongs to this process or an outlier is rather easy. The data point in question is outside the triangle.

Had we used a bell curve, the answer is not so easy; at least it will not be a straightforward reply. One would say there is chance *p* that the data point belongs to the process. Then we will apply a policy to decide whether the p value is less than the "critical value." If the p value is less, then we would say that the data point is significantly different from the process. It is a roundabout way of saying that p is outside and is certainly confusing to many who would rather have a simple and transparent answer.

The triangle presents a crystal clear view of process and facilitates straight decision making.

Review Questions

1. Why is the triangular distribution preferred in the place of uniform distribution for estimation models?
2. Can we represent data skew in a triangular model?
3. Compare the way of calculating expected value using the PERT formula with the method of calculating the expected value using the mean of triangular distribution.
4. Compare beta distribution with the triangular distribution.
5. Why is the triangular distribution popular in project management?

Exercises

1. In a triangular distribution model for productivity, $a = 30$, $b = 90$, and $c = 55$. The numbers are LOC/per day, standing for productivity in software development. The naming conventions are shown in Figure 15.2. Calculate skew.
2. Calculate mean and median productivity in the above-mentioned situation.
3. If the threshold productivity is 40, what is the risk in productivity performance in the above context?
4. What is the standard deviation of the distribution in the above example?
5. Calculate risk using a Gaussian model using the formulas given in Chapter 13 "Bell Curve," making use of the standard deviation you found in Exercise 4 and the mean you found in Exercise 2.

References

1. S. Kotz and J. R. van Dorp, *BEYOND BETA: Other Continuous Families of Distributions with Bounded Support and Applications*, World Scientific Publishing Co Pte Ltd, 2004.
2. A. S. Wahed, The family of curvi-triangular distributions, *International Journal of Statistical Sciences*, 6(special issue), 7–18, 2007.

3. M. Brizz, A skewed model combining triangular and exponential features: The two-faced distribution and its statistical properties, *Austrian Journal of Statistics*, 35(4), 455–462, 2006.
4. J. R. van Dorp and S. Kotz, A novel extension of the triangular distribution and its parameter estimation, *Journal of the Royal Statistical Society, Series D (The Statistician)*, 51(1), 63–79, March, 2002.

Chapter 16

The Law of Life: Pareto Distribution— 80/20 Aphorism

Pareto distribution is a fat-tailed skewed distribution invented by Vifredo Pareto. A brief biography of Pareto is given in Box 16.1. The distribution was originally used to describe wealth distribution in society. Larger wealth is controlled by fewer people.

> **BOX 16.1 VILFREDO PARETO—THE ECONOMIST**
> **WHO DISCOVERED MANAGEMENT (1848–1923)**
>
> Vilfredo Pareto was an Italian sociologist, engineer, economist, philosopher, political scientist, and mathematician.
>
> Between 1859 and 1864, Vilfredo changed schools several times. From 1864 to 1867, Vilfredo studied mathematics and physics at the Università di Torino.
>
> In 1869, he earned a doctor's degree in engineering from what is now the Polytechnic University of Turin. His dissertation was titled "The Fundamental Principles of Equilibrium in Solid Bodies." His later interest in equilibrium analysis in economics and sociology can be traced back to this paper.
>
> After his studies, Pareto worked for some years at the Italian Railway Company and traveled to Germany, England, Belgium, Switzerland, and Austria. In the field of statistics, Pareto worked for insurances and the calculation of pensions.

Pareto became famous by the Pareto Optimum in economics and the Pareto distribution. In 1896, he found that the distribution of income does not follow the normal distribution but is mostly inclined to the right side. His discovery of the "distribution curve for wealth and incomes" of 1895 made Pareto famous as a statistician.

The Pareto principle was named after him and built on observations of his such as that 80% of the land in Italy was owned by 20% of the population.

Pareto was the first to realize that utility was a preference ordering. With this, Pareto not only inaugurated modern microeconomics but also demolished the alliance of economics and utilitarian philosophy. Pareto said "good" cannot be measured. He replaced it with the notion of Pareto optimality, the idea that a system is enjoying maximum economic satisfaction when no one can be made better off without making someone else worse off. Pareto optimality is widely used in welfare economics and game theory. A standard theorem is that a perfectly competitive market creates distributions of wealth that are Pareto optimal.

His legacy as an economist was profound. Partly because of him, the field evolved from a branch of social philosophy as practiced by Adam Smith into a data-intensive field of scientific research and mathematical equations. (http://en.wikipedia.org/wiki/Pareto_principle; http://en.wikipedia.org/wiki/Pareto_distribution)

Structure of Pareto

Pareto is known as a fat-tailed distribution. Gaussian, exponential, and Pareto tails are compared in Box 16.2. It is shown that Pareto has the largest tail.

A graph of the Pareto distribution is plotted in Figure 16.1. The probability of usage of software features is the metric plotted in Figure 16.1. The distribution begins from its mode and extends asymptotically to the right. The decline of usage is gradual.

The Pareto probability density function (PDF) depends on two parameters, mode m and shape factor α. The equation to the PDF is shown as follows:

$$\text{PDF} = \frac{\alpha m^{\alpha}}{x^{\alpha+1}} \tag{16.1}$$

The equation can be rewritten by marking the constant term separately and bringing the variable term to the numerator, as follows:

$$f(x) = (\alpha m^{\alpha})x^{-(\alpha+1)} \tag{16.2}$$

The equation is clearly a form of the power law with a negative exponential x^{-b}. Power law is one of the favorite curves used in data mining.

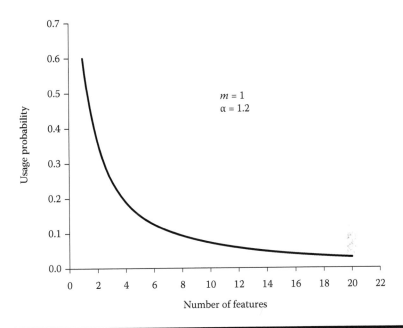

Figure 16.1   Pareto distribution of features usage.

BOX 16.2   A STORY OF TAILS

The Gaussian tail dies soon. The exponential tail stretches longer but is limited. The Pareto tail, resulting from the power law, is unlimited. We can compare the standard forms of these three tail equations:

Gaussian, standard form = $e^{-\frac{1}{2}x^2}$
Exponential, standard form = e^{-x}
Pareto, standard form = x^{-1}

In the previously mentioned expressions, scale factor = 1 and location = 0. If we check the value of tails at $x = 6$, we find

Gaussian tail = 0.0000000152
Exponential tail = 0.00248
Pareto tail = 0.167

At $x = 6$, the Gaussian tail is nearly zero, and the exponential tail is 162,755 times bigger. In turn, the Pareto tail is 67 times stronger than the exponential. For larger values of x, divergence among the three tails increases further. The Gaussian tail will be dead, the exponential tail will slide toward zero, and the Pareto tail will still have significant values for a long distance.

These three tails represent three aspects of engineering and management. Gaussian is drawn to its center; its body is accentuated and its tail attenuated, a true model of process behavior. The Gaussian tails are either process defects or rejection areas.

Exponential curve represents decay or defects in a product. There seem to be special mechanisms in a product that cause decay or vulnerabilities that cause defects. By definition, exponential tail represents failure, not performance of products.

Pareto is often a model for external factors that influence a product or a process from outside the organization.

Business comprises effects represented by these three tails.

The cumulative distribution function is shown in Figure 16.2. The y-axis directly reads usage probability, while the x-axis reads the number of features. Using this model, we can find quickly the usage probability of n number of features in a software product.

The equation to cumulative distribution is rather simple and is shown as follows:

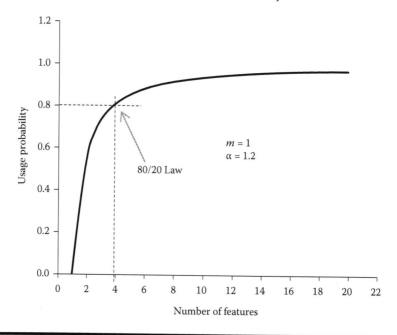

Figure 16.2 Cumulative Pareto distribution of features usage.

$$\text{CDF} = 1 - \left(\frac{m}{x}\right)^{\alpha} \tag{16.3}$$

It may be noted that the previously mentioned equations are defined for values of x greater than mode m.

Key statistics of the distribution are given as follows:

$$\text{Mean} = \frac{\alpha m}{\alpha - 1} \tag{16.4}$$

$$\text{Median} = m2^{\frac{1}{\alpha}} \tag{16.5}$$

The mean is defined for values of shape factor $\alpha > 1$.

An Example

A Pareto model has been established with mode $m = 1$ and shape factor $\alpha = 1.2$ in Data 16.1. The mean for this model turns out to be 6 while the median is 1.8. The fact that the mean is so far away from the median explains a model skew. The mean has shifted toward the tail. The PDF and cumulative distribution function (CDF) have computed and the values are shown in Data 16.1. Pareto calculations are easy and can be managed with basic Excel.

The 80/20 Law: Vital Few and Trivial Many

The CDF shown in Figure 16.2 allows us to think of the famous 80/20 due to Pareto. It may be seen that 20% of features have 80% usage probability. This is a basic principle used in statistical testing. This model is also called the operational profile of the product. There are many 80/20 laws that rule life. A brief list is given in Box 16.3.

The 80/20 law depicts the phenomenon of "vital few and trivial many." Illes-Seifert and Paech [1] have analyzed application of this principle to software defects. They report,

> The distribution of about 430 defects over about 500 modules has been analysed and confirms the Pareto Principle, i.e. approximately 80% of the defects were contained in 20% of the modules.

Data 16.1   Pareto Distribution of Usage

Parameters		
m	1	
α	1.2	
Statistics		
Mean	6	
Median	1.8	

No. of Features	Probability Density Function (PDF) of Usage	Cumulative Distribution Function (CDF) of Usage
1	0.600	0.000
2	0.364	0.565
3	0.253	0.732
4	0.191	0.811
5	0.152	0.855
6	0.125	0.884
7	0.106	0.903
8	0.091	0.918
9	0.080	0.928
10	0.071	0.937
11	0.064	0.944
12	0.058	0.949
13	0.053	0.954
14	0.049	0.958
15	0.045	0.961
16	0.042	0.964
17	0.039	0.967
18	0.036	0.969
19	0.034	0.971
20	0.032	0.973

Illes-Seifert and Paech set up four Pareto hypotheses and prove each is right.

1. Pareto distribution of defects in files: a small number of files accounts for the majority of defects
2. Pareto distribution of defects in files across releases
3. Pareto distribution of defects in code: a small part of the system's code size accounts for the majority of defects
4. Pareto distribution of defects in code across releases

In another example of Pareto, Ostrand and Weyuker [2] have studied the distribution of defects over different files in 13 releases of a large industrial inventory tracking system. For each release, the faults were always heavily concentrated in a relatively small number of files. For example, they find that in a certain release, 10% of files account for 68% of the faults. Interestingly, they find a similar pattern in code size; a small number of files contain more code.

In yet another example, Murgia et al. [3] studied the defects in two open source Java projects, both developed following agile practices, and find, "There are few Compilation Units hosting most bugs, and most other Compilation Units are with a very few bugs," supporting the 80/20 phenomenon. They report that 80% of bugs are contained in compilation units ranging from 8% to 20%; clearly, the Pareto law holds. The researchers observe, "This is an important result from the software engineering point of view. In fact, a review of a small fraction of faulty Compilation Unit may have an exponential impact on the overall amount of software defects detectable and fixable."

BOX 16.3 THE 80/20 LAWS

The 80/20 principle—that 80% of result flows from just 20% of the causes—is the one true principle of highly effective people. It has become a management law. Its effect on all facets of life may be seen from the following compilation:

Pareto's historic observations:

80% of Italy's land was owned by 20% of the population
20% of the pea pods in his garden contain 80% of the peas

In software development,

80% of errors and crashes come from 20% of bugs
20% of software components contain 80% of defects
20% of defects cause 80% of down time
20% of test cases capture 80% of defects

In problem solving,

20% of problems cause 80% of damage
20% of causes are responsible for 80% of problems
20% of hazards account for 80% of injuries
20% of customers take up 80% of one's time
80% of crimes are committed by 20% of criminals

In the Internet,

1% of the users of a website create new content, 99% lurk

In general,

20% of humans hold power over the remaining 80%
20% patients use 80% health care resources
10% of expenditure on health helps 90% of poor people
20% of the world's population controlling 82.7% of the world's income

In business,

80% of a company's profits come from 20% of its customers
80% of a company's complaints come from 20% of its customers
80% of a company's profits come from 20% of the time spent by its staff
80% of a company's sales come from 20% of its products
80% of a company's sales are made by 20% of its sales staff
80% of your sales come from 20% of your clients

Related special findings:

A small number of flows carry most Internet traffic, and the remainder
consists of a large number of flows that carry very little Internet traffic
(Elephant flow and mice flow).
The first 90% of the code accounts for the first 90% of the development
time. The remaining 10% of the code accounts for the other 90% of the
development time (Tom Cargill).
Ninety percent of everything is crap (Theodore Sturgeon).
80% of your benefits come from 20% of your efforts (Tim Ferriss in *The
4 Hour Workweek*).

Generalized Pareto Distribution

Open source projects have a different DNA. They follow the Pareto distribution.

Simmons and Dillon [4] note a Pareto distribution in the size of the number of developers participating in open source projects with most projects having only one developer and a much smaller percentage with larger, ongoing involvement.

Kolassa et al. [5] have proposed a generalized Pareto distribution to define commit sizes in open source projects.

The equation to the model is shown as follows:

$$f(x) = \frac{1}{\sigma}\left|1 + \xi\frac{x-\theta}{\sigma}\right|^{-1-\frac{1}{\xi}} \quad \text{for } \xi \neq 0 \qquad (16.6)$$

where θ is the location parameter (controls how much the distribution is shifted), σ is the scale parameter (controls the dispersion of the distribution), and ξ is the shape parameter (controls the shape).

Kolassa et al. [5] have fitted the generalized Pareto with the following parameters derived from commit size data:

ξ shape = 1.4617
θ location = 0.5
σ scale = 13.854

The previously mentioned model explains how contributions from open source developers can vary.

Duane's Model

J. T. Duane has developed a reliability growth model based on the Pareto distribution (power law), which has long been in use National Institute of Standards and Technology (NIST).

This model implies that the reliability during any specific interval can be represented by the negative exponential model:

$$\lambda_C = kt_T^{-\alpha} \tag{16.7}$$

where C is the average estimate of cumulative failure rate, t_T is the total accumulated operating hours, k is the constant representing cumulative failure rate at $t_T = 1$, and α is the improvement rate constant.

Tailing a Body

In a novel attempt, Herraiz et al. [6] fit a Pareto tail to a log-normal body with object-oriented software metrics. The Pareto distribution, known for its prominent tail, fits the larger values better while the smaller values follow the log-normal distribution. Herraiz et al. [6] called the mixture model *double Pareto distribution*. Not all OO metrics need a double Pareto. Two metrics, the number of children and the lack of cohesion in methods, are better described using a power law for the entire range of values. Three metrics, weighted methods per class, coupling between object classes, and requests for a class, are better described by a double Pareto.

In another study, Herraiz et al. [7] have studied a large quantity of open source code, approximately 700,000 C files. In particular, the following metrics were studied: source lines of code, lines of code, number of blank lines, number of comment lines, number of comments, number of C functions, McCabe's cyclomatic complexity, number of function returns, and Halstead metrics. All the metrics were found to follow a double Pareto distribution.

Pareto is the simplest distribution available. It is also the flexible and easy to adapt. The 80/20 law derived from Pareto principle are used extensively in business management, project management, software development and problem solving. It is also used in reliability modelling.

Review Questions

1. What is the meaning of 80/20 law?
2. What is the Pareto principle?
3. Why is Pareto distribution called a fat tailed distribution?
4. Which distribution has the fattest tail: Gaussian, exponential, or Pareto?
5. Give three examples of Pareto laws.

Exercises

1. Calculate the mean value of Pareto distribution if mode = 7 and shape = 4.
2. Calculate the median value for the previous case.
3. Assume the defect density distribution of a certain application follows Gaussian tail. If the threshold of defect density is six units (relative value), what is the reliability of the application? (Clue: make use of the calculation shown in Box 16.2: A Story of Tails.)
4. Assume the defect density distribution of a certain application follows exponential tail. If the threshold of defect density is six units (relative value), what is the reliability of the application? (Clue: make use of the calculation shown in Box 16.2: A Story of Tails.)
5. Assume the defect density distribution of a certain application follows Pareto tail. If the threshold of defect density is six units (relative value), what is the reliability of the application? (Clue: make use of the calculation shown in Box 16.2: A Story of Tails.)

References

1. T. Illes-Seifert and B. Paech, *The Vital Few and Trivial Many: An Empirical Analysis of the Pareto Distribution of Defects*. Software Engineering, Kaiserslautern, Germany, 2009, pp. 151–164.
2. T. J. Ostrand and E. J. Weyuker, The distribution of faults in a large industrial software system, In: *Proceedings of the 2002 ACM SIGSOFT International Symposium on Software Testing and Analysis (ISSTA)*, ACM Press, Roma, Italy, 2002, pp. 55–64.
3. A. Murgia, G. Concas, S. Pinna, R. Tonelli and I. Turnu, *Empirical Study of Software Quality Evolution in Open Source Projects Using Agile Practices.*
4. G. L. Simmons and T. S. Dillon, Towards an ontology for open source software development, *International Federation for Information Processing*, 203, 65–75, 2006.

5. C. Kolassa, D. Riehle and M. A. Salim, *A Model of the Commit Size Distribution of Open Source*. Springer, Berlin, Heidelberg, 2013.
6. I. Herraiz, D. Rodriguez and R. Harrison, On the statistical distribution of object-oriented system properties. In: *Third International Workshop on Emerging Trends in Software Metrics (WETSoM 2012)*, [Version 20101126r], Zurich, Switzerland, June 3, 2012.
7. I. Herraiz, J. M. Gonzalez-Barahona and G. Roble, Towards a theoretical model for software growth. In: *2013 10th Working Conference on Mining Software Repositories (MSR)*, 2007, p. 21.

TAILED DISTRIBUTIONS

Tailed distributions occupy a special position in pattern recognition. They are used to describe extreme events. The representation of such less likely events is not often the primary interest of project managers and engineers. Attention to central tendencies and overall generic expressions of dispersion have dominated managerial thinking. The prediction of tails is a specialist domain, not a generalist's credo. In recent years, interest in tails has increased. Software buyers would like to estimate residual defects. Business managers would like to estimate scope creep. Engineering managers would like to predict size growth. Hence, models of software evolution were created.

The following five chapters are devoted to the use of a few tailed distributions: log-normal, gamma, Weibull, Gumbel, and Gompertz. Each distribution has unique characteristics that entail unique applications. Together, these five distributions can handle most extreme events in software engineering. A remarkable application of such distributions is in the construction of software reliability growth models, described in Chapters 19 and 21.

Despite the mathematical form, which might dissuade a casual reader, these distributions are simple to use; they are widely used in the industry. Computations needed for solving these expressions are minimal and can be accomplished using MS Excel.

Chapter 17

Software Size Growth: Log-Normal Distribution

Log-Normal Processes

Software grows in the development cycle. Software metrics, namely, size, effort, defects, and reliability, all manifest growth of software.

Growth is a multiplicative process. This sharply contrasts with the additive process of gambling.

Growth of sites on the Web, growth of organisms in biology and ecology or growth of fatigue cracks in semiconductors, growth of Web pages in the Internet, growth of pollutants in the atmosphere, growth of cancer in people, growth of corrosion in metals, growth of phone traffic in a communication network, and growth of words are examples of multiplicative processes.

Growth is inadequately represented in the bell curve. A different curve, namely, the log-normal distribution, first used in 1836 (see Box 17.1), is seen to represent adequately well such growth events. The log-normal distribution is built on a simple premise that logarithms of skewed data will be normally distributed without skew; taking logarithms removes skew. The logarithmic scale is often used to present complex nonlinear data in a simplified linear form (see Box 17.2). Logarithmic transformation of observations allows us to apply the familiar properties of the bell curve to the transformed data.

Consider software design complexity, which is relatively skewed when compared with the bell curve. We analyzed the NASA data [1] on module design complexities of 505 modules written in C language.

BOX 17.1 THE FIRST APPEARANCE OF LOG-NORMAL

It began with geometric mean. If a variable can be thought of as the multiplicative product of some positive independent random variables, then it could be modeled as log-normal.

The basic properties of log-normal distribution were discussed long ago in 1836 by Weber [2]. McAlister described the log-normal distribution around 1879. Kapteyn and Van Uven, in 1916, gave a graphical method of estimating the parameters; the log-normal distribution was found to be accurately representing the distribution of critical dose for several drugs; this was also the first time that log-normal distribution was applied in real life.

NASA software defect data sets have been made publicly available and extensively used by researchers.

The data may be viewed in the box plot provided in Figure 17.1.

In the box plot, it may be seen that design complexity data are right skewed with several outliers too. (Box plot is a good data visualizer; it produces a rich picture of data. Further information about the reading a box plot is available in Chapter 4.)

Having seen the picture of raw data, we can choose to take logarithms of data and examine the result to see how data have been transformed, in particular, how well data have been unskewed, by logarithms. We plot the logarithms of data in a box plot in Figure 17.2.

The new box plot has noteworthy and serious differences. Comparing Figure 17.2 with Figure 17.1 reveals two consequences of the transformation:

1. The box has become symmetric.
2. There is a drastic reduction in the number of outliers.

The new box plot is a better fit. It is as if data after transformation have found its destination pattern.

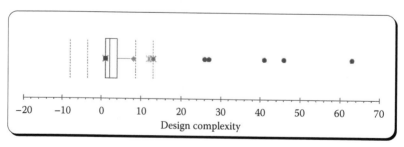

Figure 17.1 Design complexity data.

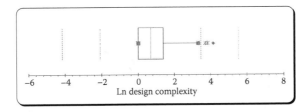

Figure 17.2   Natural logarithms of design complexity data.

The two box plot patterns can be expressed as mathematical curves. First, we can fit the data to a normal distribution. The symmetrical normal distribution is rather a mechanically executed force fit. The data have the following normal parameters:

Mean = 3.592
Standard deviation = 5.447

To obtain the parameters for log-normal distribution, in the most commonly used format, we must estimate the mean and standard deviations of natural logarithms of data. Thus, we obtain 0.771 and 0.896.

Normal and log-normal curves generated by Excel functions NORM.DIST and LOGNORM.DIST based on the two sets of above parameters are plotted in Figure 17.3.

The normal curve represents a traditional and habitual treatment, and the log-normal curve represents a modern and theory-driven treatment of design complexity data. It may be seen that the log-normal curve is in closer agreement with the box plot of data shown in Figure 17.1.

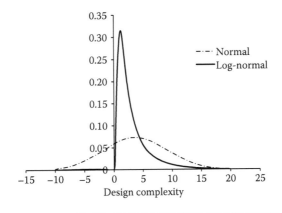

Figure 17.3   Normal versus log-normal distributions of design complexity.

The log-normal curve represents an engineering truth missed by the bell curve.

The normal curve is a misfit; it has an odd negative tail that is not practical and also it tenders a misleading peak. The log-normal curve does not go negative and has the right skewed tail and a perfect peak.

Software design is not a Gaussian process; it is a log-normal process.

Building a Log-Normal PDF for Software Design Complexity

As the saying goes, if we substitute natural logarithms for *x* in a Gaussian PDF equation (Equation 17.1), we obtain a log-normal PDF equation, as follows:

$$F(x,\mu,\sigma) = \frac{1}{\sqrt{2\pi}\sigma} e^{-\frac{(x-\mu)^2}{2\sigma^2}} \qquad (17.1)$$

Accordingly, natural logarithms of data must be taken first, and then in Equation 17.1, *x* must be replaced by Ln(*x*), μ must be replaced by the average of Ln(*x*), and σ must be replaced by the standard deviation of Ln(*x*). However, the equation will be in the logarithmic scale. Taking exponential of the results will convert them to real-life units.

It may be noted that the Excel function LOGNORM.DIST takes inputs in the logarithmic scale but gives results in real domain. We do not have to take exponentials and go through a separate conversion process, and this is a great practical convenience.

Users have built their own versions of the log-normal PDF. The first choice we need to make is the central value.

The log-normal PDF is built around the median, like the Gaussian is built around the mean.

Some versions of log-normal use the geometric mean. In a typical log-normal process, the median and the geometric mean are nearly equal. Using the geometric mean is highly justified by the fact that log-normal numbers are multiplicative and tend to form a geometric series. In creating a log-normal PDF, the *NIST Engineering Statistics Handbook* [3] proposes the following structure:

$$f(x) = \frac{1}{\sigma x\sqrt{2\pi}} e^{-\frac{(\ln x-\beta)^2}{2\sigma^2}} \qquad (17.2)$$

where β is the scale factor and σ is the shape factor.

The parameters α and β can be extracted by (1) the method of moments (MOM), (2) the maximum likelihood method, and (3) the minimum χ^2 method. We would pursue the MOM in this chapter; hence, we use the following two relationships:

$$\beta \text{ is the scale factor} = \text{mean of } Ln(x) \tag{17.3}$$

$$\sigma \text{ is the shape factor} = \text{standard deviation of } Ln(x) \tag{17.4}$$

These relationships are inherent in the Excel function LOGNORM.DIST, as we have seen while creating Figure 17.3. Methods 2 and 3 compute parameters by iteration, and it is a good idea to use Equations 17.3 and 17.4 to generate initial values that may help the following iteration runs to converge faster.

Even manual techniques of parameter extraction begin with Equations 17.3 and 17.4. If we apply them to design complexity data, the scale and shape parameters would become 0.771 and 0.896, the starting values.

The NIST suggestion becomes a valuable option: the scale parameter may be taken as $Ln(Median(x))$ instead of Mean of $Ln(x)$. The scale parameter by NIST option will be 0.693 instead of the standard 0.771. This is based on a logic that lognormal distributions are centered on the median, and we need not search for the scale parameter iteratively.

Working with a Pictorial Approach

Let us now consider a graphical way of connecting with mathematical distribution. We can construct and use a histogram, known for its pattern extraction capabilities. Such a histogram of design complexity data is shown in Figure 17.4.

The histogram has extracted a distinctive pattern, with well-defined and clearly discernible features: mode (peak), shape, and tail. These graphical features provide guidance in the choice of a sensitive log-normal parameter: the shape factor. Using graphical matching, we can select the most appropriate from a set of design complexity log-normal curves.

A set of log-normal curves are given in Figure 17.5, with four sets of log-normal parameters given as follows:

1. Shape 0.7, scale 0.5
2. Shape 0.896, scale 0.771 (obtained by MOM)
3. Shape 1.1, scale 1.0
4. Shape 1.3, scale 1.4

These curves have been obtained iteratively by perturbing the parameter values around an initial value, a second pair of parameters, with a shape of 0.896 and a scale of 0.771, obtained using MOM.

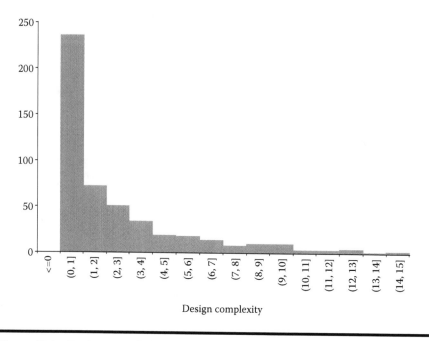

Figure 17.4   Design complexity histogram.

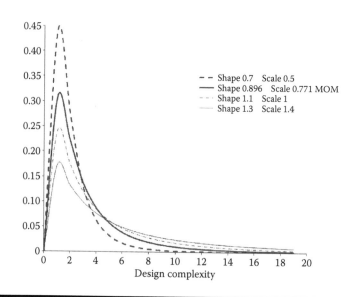

Figure 17.5   Pertubations of log-normal distribution of design complexity.

It may be seen in Figure 17.5 that the curve with MOM parameters matches the histogram, with a tail finishing at 15. The other curves either stop up front or overshoot.

It may also be seen that we have made no attempt to make the log-normal model represent the outliers seen in the box plot seen in Figure 17.1. Our emphasis has remained on the body of the log-normal and not on its tail. That emphasis depends on our strategy in building the body or the tail. Had we wished to emphasize the tail, then we would have opted for attaching a Pareto tail. To pursue this idea, we need to improve the precision of our judgment. We do so by evaluating errors in prediction. For nine values of percentiles, ranging from 0.1 to 0.9 in steps of 0.1, data values are computed first by using the percentile function in Excel. The same percentiles are interpreted as probabilities in cumulative log-normal distribution, and we predict design complexities corresponding to the nine percentiles by doing an inverse calculation using LOGNORMINV in Excel. The difference between predicted value and data value is taken as error in prediction. To find a meaningful average error, we find absolute error each time, remove the sign, and then take the average. Prediction errors are calculated for all the candidate log-normal models. The calculations are shown in Table 17.1.

In Model A, the parameters directly obtained by MOM are used. In Model B, the scale is estimated with reference to the median. In Model C, the scale is the same as Model B, but the shape has been perturbed till mean absolute error (MAE) converged to a minimum.

For further discussion, let us choose the optimized Model C having minimum error. The optimized model has a scale of 0.693 and a shape of 0.930. We can plot the log-normal cumulative distribution function (CDF) of Model C using the Excel function, as follows:

CDF = LOGNORM.DIST(*x*, scale, shape, 1).

After substitution, the expression becomes

CDF = LOGNORM.DIST(*x*, 0.693, 0.930, 1).

For our reference, we can plot the probability density function (PDF) using the Excel function, as follows:

PDF = LOGNORM.DIST(*x*, scale, shape, 0).

After substitution the expression becomes

PDF = LOGNORM.DIST(*x*, 0.693, 0.930, 0).

In both the cases, *x* is a design complexity. The plots are shown in Figure 17.6.

Table 17.1 Prediction Errors

Percentile	Data	Models												
		A		B		X1		C		X2		X3		X4
		Shape 0.896		0.896		0.92		0.93		0.94		0.93		0.93
		Scale 0.771		0.693		0.693		0.693		0.693		0.68		0.7
		Prediction	Absolute Error	Prediction	Absolute Error	Prediction	Absolute Error	Prediction	Absolute Error	Prediction	Absolute Error	Prediction	Absolute Error	Prediction	Absolute Error
0.1	1	0.7	0.314	0.6	0.051	0.6	0.019	0.6	0.008	0.6	0.008	0.6	0.000	0.6	0.012
0.2	1	1.0	0.017	0.9	0.076	0.9	0.019	0.9	0.008	0.9	0.008	0.9	0.004	0.9	0.018
0.3	1	1.4	0.351	1.2	0.101	1.2	0.016	1.2	0.006	1.2	0.006	1.2	0.009	1.2	0.024
0.4	1	1.7	0.723	1.6	0.129	1.6	0.010	1.6	0.004	1.6	0.004	1.6	0.016	1.6	0.032
0.5	2	2.2	0.162	2.0	0.162	2.0	0.000	2.0	0.000	2.0	0.000	2.0	0.026	2.0	0.040
0.6	2	2.7	0.713	2.5	0.204	2.5	0.015	2.5	0.006	2.5	0.006	2.5	0.039	2.5	0.050
0.7	3	3.5	0.459	3.2	0.260	3.2	0.041	3.3	0.017	3.3	0.017	3.2	0.059	3.3	0.065
0.8	5	4.6	0.404	4.3	0.345	4.3	0.087	4.4	0.037	4.4	0.037	4.3	0.093	4.4	0.087
0.9	8	6.8	1.184	6.3	0.511	6.5	0.197	6.6	0.084	6.7	0.085	6.5	0.170	6.6	0.131
MAE		0.48082		0.20445		0.04475		0.01889		0.01903		0.04641		0.05113	

Note: MAE, mean absolute error, obtained from three important log-normal models given as follows: (A) Scale = Ln (mean), 0.771; Shape = SD Ln(x) 0.896; MAE = 0.481; (B) Scale = Ln (median), 0.693; Shape = SD Ln(x) 0.896; MAE = 0.204; (C) Scale = Ln (median), 0.693; Shape = 0.930; MAE = 0.019.

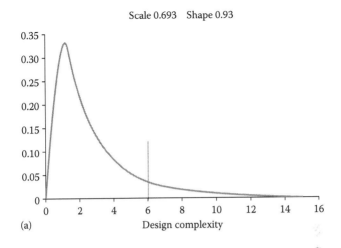

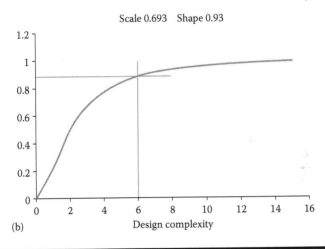

Figure 17.6 Optimal log-normal (a) PDF and (b) CDF of design complexity.

BOX 17.2   LOGARITHMIC SCALE

The human ear responds to sound in a logarithmic scale; the response has an amazing dynamic range, from extremely small to extremely large sound. Such a range is possible if the scale were logarithmic. If the sound level increases tenfold, the response increases one notch. Sound level is measured in decibels, logarithms to the base 10 of sound intensities.

Earthquakes are measured in a logarithmic scale. If the power unleashed is 10 times larger in the Richter scale (invented by Charles F. Richter in 1934) for measuring earthquake strength, the signal jumps one point. The release of a million-fold strong outburst appears as a mere six-point movement in the Richter scale. At the same time, the Richter scale is sensitive enough to register very small seismic activities, too small to be detected by humans. The range of the Richter scale is enormous.

Orders of magnitude are seen through logarithms. The conversion from logarithms to real scale is achieved by taking antilogarithms. A trained human mind quickly does a conversion by applying rules and examples. In this context, the log-normal distribution is a return to a natural way of dealing with huge magnitudes. One just has to get used to it.

Application of the Log-Normal Model 1

If an engineering limit on design complexity is set at 6, then we can find the probability of meeting this limit (certainty) from the CDF. In the CDF shown in Figure 17.6, a vertical line runs from $x = 6$ to meet the CDF; from the point of intersection, a horizontal line is drawn, which meets the y axis at approximately 0.9.

The exact value is obtained from the CDF as follows:
Probability that design complexity is <6:

= CDF ($x = 6$)
= LOGNORM.DIST (6,0.693,0.930,1)
= 0.8813

The above number represents the certainty of meeting the design complexity goal of 6.

We can extend the analysis to calculate the risk of design complexity exceeding the limit of 6.

The risk of design complexity exceeding the limit of 6 is as follows:

= 1 − CDF ($x = 6$)
= 1 − LOGNORM.DIST (6,0.693,0.930,1)
= 1 − 0.8813
= 0.1187

Calculating certainty and risk is a most useful application of log-normal distribution.

Application of the Log-Normal Model 2

The next application refers to developing a control chart for design complexity. We cannot apply the Shewhart Control Chart with the mean μ as the central line and μ ± 3 σ as the control limits. Shewhart chart assumes a normal distribution, and its design complexity is log-normal. Shewhart limits are symmetrical design complexity limits that cannot be symmetrical. We have established design complexity as a skewed distribution. Besides, we do not need a lower control limit for complexity. All we need is an upper control limit.

Shewhart limits include 99.73% of process inside the limits and keep only 0.27% as outliers. Shewhart limits apply better for manufacturing processes. For creative processes such as software design, the authors suggest a different rule for control limits. The proposed limits include 95% (0.95) of processes inside the limits and mark 5% of processes as outliers. This upper control limit is obtained from the CDF shown in Figure 17.6 as an *x* value corresponding to a *y* value of 0.95.

Upper control limit for *y* = 0.95
= LOGNORMINV (*y*, scale, shape)
= LOGNORMINV (0.95,0.693,0.93)
= 9.2324

This sets the statistical limit for upper control of design complexity.

Features Addition in Software Enhancement

In Chapter 13, we treated requirement volatility to a Gaussian with a standard deviation of approximately 3.3% and a mean value of 4%, for full life cycle development projects with stringent business control on requirements. In large enhancement projects, the Gaussian model does not hold; here changes are far more common. Features added after requirements are "finalized" can touch high values, as high as 50%. The growth of features is log-normal. The pattern of growth varies. Three examples, A, B, and C, are presented in Figure 17.7.

Model C has the largest scale of 20 and the fattest tail. Model B has a scale of 10 and a medium-sized tail. Model A has a scale of 4 and has an early finishing point. All three models represent the customer's processes over which the maintenance team has no direct control. In such cases, statistical management reduces to empirical understanding of the process with data and creating the appropriate PDFs. To recognize if variation is Gaussian or log-normal is the first step; this is enabled by histograms. Fitting the appropriate PDF by parameter extraction is the next step. Applying the model to solve problems and take decisions is the goal.

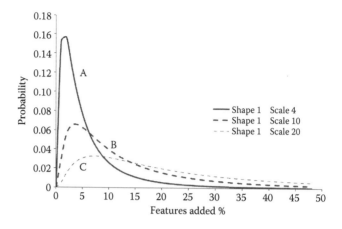

Figure 17.7   Log-normal PDF of software enhancements.

A Log-Normal PDF for Change Requests

The process of "change requests" in a support project is a mixture; it is composed of assorted tasks, including bug fix, feature addition, and patchwork. A PDF of change requests is modeled with the following parameters:

Shape σ = 1
Scale β = 7

The scale factor is set at the median of data. The shape factor is chosen by an iterative search for best fit. The log-normal PDF for change requests is plotted in Figure 17.8.

However, this is merely curve fitting. This model does not benefit from the ideological context such as that present in the model for feature addition or design complexity. Despite this limitation, the model can still be used for forecasting.

A better approach, beyond the scope of this book, would be to create a mixture model, combining inherent probabilistic characteristics of the components.

> *Bug fixes may be denoted by Weibull distribution, patchwork by beta distribution and feature addition by log-normal distribution.*

Mathematically combining the three would require a series of approximations and special analytical treatments, which would require a specialist's knowledge. However, such a combination can also be achieved digitally by simulation.

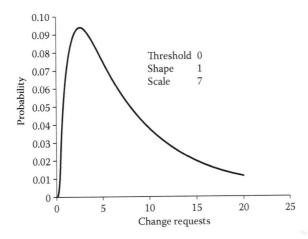

Figure 17.8 Log-normal PDF of change requests.

From Pareto to Log-Normal

Pareto distribution is a power law and is known for its fat tail (see Chapter 16 for more details). Log-normal is a growth model and also has a limited-sized tail. One can switch to Pareto if bigger tails are needed. However, the similarity does not end in tails.

> When examining income distribution data, Aitchison and Brown (1954) observe that for lower incomes a lognormal distribution appears a better fit, while for higher incomes a power law distribution appears better [4].

Power law distributions and log-normal distributions are quite natural models and can be generated intuitively. They are also intrinsically connected.

Power law and log-normal both have been applied to file size distributions in the Internet; they fare equally well. From a pragmatic point of view, it might be reasonable to use whichever distribution makes it easier to obtain results.

Some Properties of Log-Normal Distribution

Some properties of log-normal distribution can come in handy while analyzing data. The following median-related formulas are given in NIST:

$$\text{Mean} = \text{Median } e^{\frac{\sigma^2}{2}} \qquad (17.5)$$

$$\text{Variance} = \text{Median}^2 \, e^{\sigma^2} \left(e^{\sigma^2} - 1 \right) \tag{17.6}$$

The central tendencies are defined in terms of parameters as follows:

$$\text{Mean} = e^{\left(\beta + \frac{\sigma^2}{2} \right)} \tag{17.7}$$

$$\text{Median} = e^{\beta} \tag{17.8}$$

$$\text{Mode} = e^{\left(\beta - \sigma^2 \right)} \tag{17.9}$$

We can see from the previous equations that the mean is always larger than the median. Similarly, the mode is the smallest.

When β = 1, the log-normal distribution is called *standard log-normal distribution*.

Case Study—Analysis of Failure Interval

Log-normal distribution is widely used in reliability studies. NIST presents several models for reliability analysis, and log-normal is one of them. The choice depends on interpretation of the famous bathtub curve. Initially, mechanical systems show infant mortality with a failure rate that increases till the system stabilizes. Then failure rate decreases and reaches a flat low level. When the failure rate is constant, the exponential distribution is enough.

When the failure rate is changing, log-normal or Weibull or other models capable of handling change are required.

Reliability Analysis Centre [5] illustrates an example of log-normal distribution with a scale of 10.3 and a shape of 1.0196 to represent infant mortality and speedy recovery, although Weibull is their favorite model for reliability analysis.

> *In mechanical systems reliability decreases with time whereas in software products reliability increases with usage, bug discovery and bug fixing.*

Failure mechanisms propagate and grow in physical systems; in software, they are located, confined, and eliminated. We need to bear this in mind while working on developing a probabilistic model for software reliability.

Failure models also use theory of product and ensure relevance. For example, Varde [6] developed a log-normal model based on physics of failure involving electromigration. Varde, ardently supporting physics based reasoning and apparently reluctant to use of mindless statistical models, observed,

> Nevertheless, statistics still forms the part of physics-of-failure approach. This is because prediction of time to failure is still modeled employing probability distribution. Traditionally log-normal failure distribution has been used to estimate failure time due to electromigration related failure.

Varde used median time to fail as the scale parameter and standard deviation as the shape parameter, exactly as in NIST guidelines.

We have studied failure times of software after release, the data made available by the Cyber Security and Information Systems Information Analysis Center CSIAC [7]. CSIAC is a Department of Defense (DoD) Information Analysis Center (IAC) sponsored by the Defense Technical Information Center (DTIC). The CSIAC is a consolidation of three predecessor IACs: the Data and Analysis Center for Software (DACS), the Information Assurance Technology IAC (IATAC) and the Modeling and Simulation IAC (MSIAC), with the addition of the Knowledge Management and Information Sharing technical area.

The software reliability data set has 111 records of failure intervals. With time, the failure intervals grow, increasing software reliability. We consider time between failures (TBF) as the key indicator of a complex process involving usage and maintenance. Growth of TBF is expected with a smooth log-normal with a clear peak and a distinct tail (see Box 17.3 for an analogy for software TBF).

However, the histogram of TBF, shown in Figure 17.9, reveals two peaks, belonging to two separate clusters, suggesting two growth processes. It could be that the second cluster could arise from a second release; it could also arise from a new pattern of usage recently introduced.

We have fitted two log-normal curves to the clusters. The first has a scale of 15.5, Ln(median), and a shape of 0.8 (standard deviation of Ln(x)). The second has a scale of 16.4 and a shape of 0.1. The graphs are shown in Figure 17.10. This is a composite model.

The second log-normal curve in Figure 17.10 resembles Gaussian, but still we prefer the log-normal equation because it is median based.

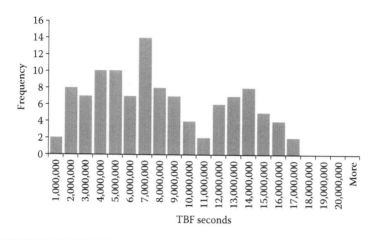

Figure 17.9   Histogram of TBF.

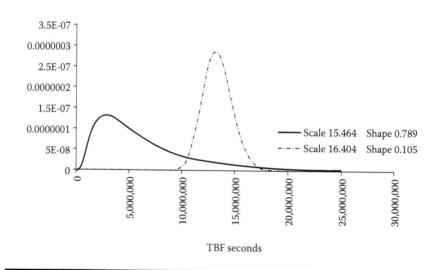

Figure 17.10   Log-normal PDF of TBF clusters.

BOX 17.3 ANALOGY—ACCELERATED LIFE TEST

If we record time served by computer hardware before the first failure occurs, what we obtain is life data, and the distribution is called *life distribution*. Life data manifest log-normal distribution, both for machines and for humans. Log-normal distribution for machine failure is used to measure and improve reliability. Log-normal distribution for human life is used to calculate life insurance premiums. In both the cases, we estimate or "measure" life properties using log-normal distribution.

In accelerated life tests of systems, extreme conditions are created, mimicking real-world scenarios, to stimulate failure events much earlier than normal. Moreover, the tests are stopped either after a certain time or after a certain number of failures occur. Tests are neither conducted indefinitely nor till all failures occur. Life data thus obtained are truncated or "censored." The picture obtained is partial, but the full picture can be constructed by statistical analysis. One such attempt is to fit log-normal distribution to censored life data and see even the unseen part of the full story of failure probabilities. Accelerated life tests are faster and cheaper.

Dube et al. [8] discussed the problem of applying log-normal distribution to censored life data, particularly parameter extraction. They analyze car failure data from Lawless [9] for this purpose. The Lawless data "shows the number of thousand miles at which different locomotive controls failed in a life test involving 96 controls. The test was terminated after 135,000 miles, by which time 37 failures had occurred."

Dube et al. took up and answered the question whether the data fit the log-normal distribution or not. They showed that "the data fits reasonably well."

Analogically, software is stressed by usage testing (e.g., user acceptance testing), triggering failure events. A record of failure times is called *life data*. Tests are not indefinitely conducted but terminated at some point of time, either after a certain number of defects have been found or after the lapse of certain time, depending on estimation and strategy of testing. Life data thus obtained can be fitted to log-normal.

Review Questions

1. Compare normal distribution with log-normal distribution.
2. Provide an example of the logarithmic scale used in practice.
3. What is the formula for the mean of log-normal distribution with shape σ and scale β?
4. What is the formula for the variance of log-normal distribution with shape σ and scale β?
5. Who invented the log-normal distribution?

Exercises

1. Download reliability data from CSIAC website [7]. Select postrelease failure events for any one project. Draw a histogram of the failure interval data. Fit the failure interval data to a log-normal distribution.
2. Use the above model to predict the current reliability related to the software project data you have selected.

References

1. Available at http://nasa-softwaredefectdatasets.wikispaces.com.
2. E. Limpert, W. A. Stahel and M. Abbt, Log-normal distributions across the sciences: Keys and clues, *BioScience*, 51(5), 341–352, 2001.
3. *NIST/SEMATECH Engineering Statistics Handbook*. The National Institute of Standards and Technology (NIST) is an agency of the U.S. Department of Commerce. Available at http://www.nist.gov/itl/sed/gsg/handbook_project.cfm.
4. J. Aitchison and J. A. C. Brown, On criteria for descriptions of income distribution, *Metroeconomica*, 6, 88–107, 1954.
5. *Journal of the Reliability Analysis Center*, Volume 13, Second Quarter, RAC, New York, 2005.
6. P. V. Varde, Role of statistical vis-à-vis physics of failure methods in reliability engineering, *Journal of Reliability and Statistical Studies*, 2(1), 41–51, 2009.
7. Available at https://sw.thecsiac.com/databases/sled/swrelg.php.
8. S. Dube, B. Pradhan and D. Kundu, Parameter estimation of the hybrid censored log-normal distribution, *Journal of Statistical Computation and Simulation*, 81(3), 275–287, 2011.
9. J. F. Lawless, *Statistical Models and Methods for Lifetime Data*, Wiley, New York, 2003.

Chapter 18

Gamma Distribution: Making Use of Minimal Data

Gamma distribution is a more general version of the exponential distribution. It provides all the advantages of the exponential distribution: it can model arrival times, and it has a fat tail and can characterize failure data. Gamma distribution has the extra advantage: it provides us a prominent mode and gives us the freedom to set the mode wherever we want by adjusting the shape factor. Gamma distribution retains the fat tail of the exponential distribution. This is not surprising because gamma distribution can be proven as a sum of exponential distributions.

The gamma distribution has two parameters, shape parameter α and scale parameter β. The probability density function (PDF) is given by the following:

$$G(x) = \frac{1}{\beta^{\alpha}\Gamma(\alpha)} x^{\alpha-1} e^{-\frac{x}{\beta}} \quad \alpha, \beta > 0 \tag{18.1}$$

where α is the shape parameter, β is the scale parameter, and $\Gamma(\alpha) = (\alpha - 1)!$ (gamma function).

In the previously mentioned PDF, the symbol $\Gamma(\alpha)$ stands for the gamma function. The PDF is plotted in Figure 18.1 to show how the shape of the distribution changes when we change the value of shape parameter from 1.2 to 2 and 3 in the plots. The scale parameter is kept constant at 10.

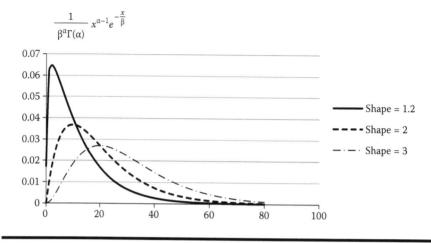

Figure 18.1 Gamma distribution.

The plots have been made using the Excel function GAMMADIST. This function returns both the PDF and the Cumulative Distribution Function CDF.

The Excel syntax is defined as follows:

$$PDF(x) = GAMMA.DIST\ (x, \text{shape parameter, scale parameter, 0})$$

$$CDF(x) = GAMMA.DIST\ (x, \text{shape parameter, scale parameter, 1})$$

BOX 18.1 SIMILARITY BETWEEN GAMMA AND LOG-NORMAL: MAKING THE CHOICE

Gamma and log-normal distributions look alike.

Kundu and Manglick [1] compared gamma and log-normal distribution and found them remarkably similar. They have used Lawless [2] data of bearing failure for this study. Let us develop some ideas around this analysis.

For Lawless data, we obtain gamma shape = 3.7138 and scale = 19.4489, relating to Equation 18.1. Using Excel GAMMA.DIST (x, shape, scale, 0), the gamma curve can be realized.

For the same data, we can obtain logarithms and find log-normal parameters, mean of natural logarithms, and standard deviation of natural logarithms of data. The log-normal curve can be realized by using Excel LOGNORM. DIST (x, mean of Ln, standard deviation of Ln, o).

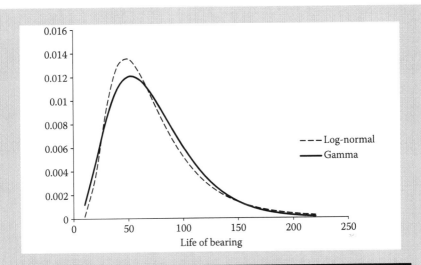

Figure 18.2   Comparison of gamma and log-normal distributions of bearing life.

Both the curves are shown in Figure 18.2.

The curves look similar. Kundu and Manglick have used the maximum likelihood estimation (MLE) technique to obtain parameters, and they obtain slightly different values but nearly identical distributions. (The Kolmogorov–Smirnov (K-S) distance between the fitted empirical distribution function and the fitted log-normal distribution function is 0.09, and the K-S distance between the fitted empirical distribution function and the fitted gamma distribution function is 0.12.)

These distributions are close to one another, and the log-normal is nearer to empirical data based on the K-S distance analysis.

The similarity is superficial. There is a difference in the approach and assumptions in constructing both the distributions.

Hence, we face the question, which distribution should be used? Are there preferences?

> *The gamma distribution may be used while taking shape based decisions by expert judgment of shapes and mean values.*
> *Log-normal distribution may be used for more rigorous numerical treatment based on parameter extraction from data alone.*

Gamma distribution has a definite advantage: it can quickly convert visual judgment to a mathematical model.

Gamma Curves for Clarification Time Data

We can model clarification time data with gamma distribution. In software maintenance, clarification time depends mostly on the customer and is not under the direct influence of the project team. Let us consider data with descriptive statistics shown in Data 18.1.

Data 18.1 shows that the mean is 36.1622, the mode is 11, and the standard deviation is 39.3173.

We wish to mention two properties of gamma distribution,

$$\text{Mean} = \text{scale} \times \text{shape}$$
$$= \alpha\beta \tag{18.2}$$

$$\text{Variance} = \alpha\beta^2 \tag{18.3}$$

To select the shape parameter, let us consult the histogram of clarification time data, as shown in Figure 18.3.

The shape of the histogram is closer to the first curve in Figure 18.1, with a shape factor of 1.2.

Substituting the values of mean (36.1622) and shape (1.2) in Equation 18.1, we obtain the value of scale as follows:

$$\text{Scale} = \frac{\text{Mean}}{\text{Shape}} = \frac{36.1622}{1.2} = 30.1352$$

Data 18.1 Descriptive Statistics of Clarification Time Data

Clarification Days	
Mean	36.16227
Standard error	4.451811
Median	21.30242
Mode	11
Standard deviation	39.31733
Sample variance	1545.852
Kurtosis	2.071553
Skewness	1.533407
Range	171.3028
Minimum	−7.30285
Maximum	164
Sum	2820.657
Count	78

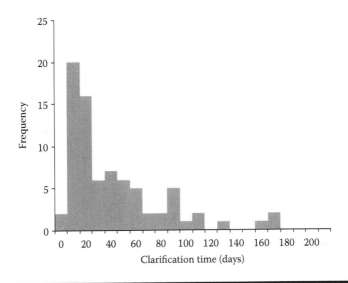

Figure 18.3   Histogram of clarification time.

A gamma distribution is fitted to the data with a shape of 1.2 and a scale of 30.1352. A plot of the fitted gamma PDF is shown in Figure 18.4.

This is the model for clarification time in software maintenance. From the model, one can make several judgments, including the following:

The PDF ends practically at 150. Therefore, the data cluster beyond 150 represents extreme values or outliers. A root cause analysis must be conducted for this excessive delay, and corrective and preventive action must be initiated.

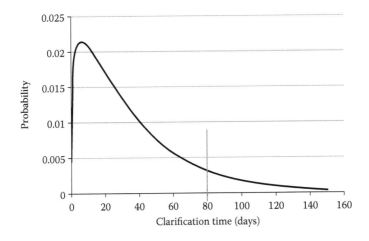

Figure 18.4   Gamma distribution of clarification time with a shape of 1.2 and a scale of 30.1352.

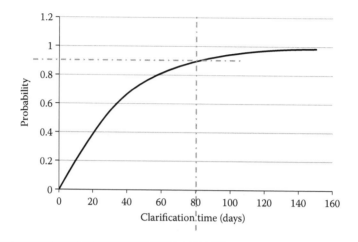

Figure 18.5 Gamma cumulative distribution of clarification time with a shape of 1.2 and a scale of 30.1352.

If there is a tolerance limit for clarification time set at 80 days by the maintenance team, then we can mark a line at $x = 80$ at the PDF. This line defines a tail, whose area represents risk. This line is marked in Figure 18.4. This is a fat tail indeed, indicating high risk. We have chosen gamma distribution to produce this fat tail and capture the hidden risk loud and clear.

To judge risk, we better plot the cumulative gamma distribution for the same scale and shape parameters. This CDF is shown in Figure 18.5.

The line from the tolerance limit of 80 days meets the CDF, and from the meeting point, a horizontal line is drawn, which meets the y axis at around 0.9. The y axis represents cumulative probability, and 0.9 means that there is a 90% chance of clarification time being within the limit. The risk of exceeding the limit is therefore 10%.

Shifting the Gamma PDF

Assume that the customer specifies a minimum time they need to resolve this problem, given the fact that the related managers are constantly traveling and communication with them slows down. Assume further that the customer specified a minimum of 4 days. Building this minimum time into the gamma function means "shifting of the curve" to the right by 4 units of time, as shown in Figure 18.4. That means that the location of the curve is shifting from $x = 0$ to $x = 4$. This minimum defines the location parameter μ.

With the inclusion of a location parameter μ, the gamma PDF equation experiences a change. The change is realized by substituting in Equation 18.1 $(x - \mu)$ in the place of x. After the inclusion of μ, the new equation of the shifted gamma PDF is given as follows:

$$G(x) = \frac{1}{\beta^{\alpha}\Gamma(\alpha)}(x-\mu)^{\alpha-1}e^{-\frac{x-\mu}{\beta}} \quad x > \mu;\ \alpha, \beta > 0 \tag{18.4}$$

BOX 18.2   INVENTOR OF GAMMA DISTRIBUTION

Leonhard Euler (1707–1783), one of the greatest mathematicians of all time, is credited with the discovery of gamma function. Some say his teacher, Bernoulli, another mathematician, invented it first.

Euler was born in Switzerland, in the town of Basel. At 13 years of age, Euler was already attending lectures at the local university. In 1723, he gained his master's degree, with a dissertation comparing the natural philosophy systems of Newton and Descartes. He wrote two articles on reverse trajectory, which were highly valued by his teacher Bernoulli.

At this time, a new center of science had appeared in Europe—the Petersburg Academy of Sciences. As Russia had few scientists of its own, many foreigners were invited to work at this center, among them Euler. On May 24, 1727, Euler arrived in Petersburg.

Euler took a very active role in the observation of the movement of Venus across the face of the sun, although at this time he was nearly blind. He had already lost one eye in the course of an experiment on light diffraction in 1738, and an eye disease and botched operation in 1771 led to an almost total loss of vision.

However, this did not stop Euler's creative output. Until his death in 1783, the academy was presented with more than 500 of his works. The academy continued to publish them for another half century after the death of the great scientist. To this day, his theories are studied and taught, and his incredibly diverse works make him one of the founding fathers of modern science.

Generating Clarification Time Scenarios with Gamma PDF Built from Minimal Data

The three-parameter gamma function in Equation 18.4 retains the properties of the two-parameter version. Equations 18.2 and 18.3 are still relevant.

Let us try to use the gamma PDF defined in Equation 18.4 to model clarification time by the customer for three possible scenarios in a maintenance project.

Our knowledge of existing pattern and the gamma parameters we have derived from existing data are very relevant clues for this model.

Let us begin with an assurance given by the customer to reduce the mean clarification time from the current 36.16 days to 20 days. The customer has already declared that he needs a minimum of 4 days for clarification. These two numbers, 4 and 20, represent the two agreed performance levels as declared by the customer, the minimum and the mean. These are really minimal data gathered to characterize clarification time. Gamma distribution will do the rest and fit behavioral details into the model based on known patterns.

Where data are minimal, gamma distribution fills the gap.

The minimum value 4 represents the location parameter, a fixed value in the models we are going to build.

We construct three types of customer responses defined by gamma with three values for shape factors: 1.2, 2.0, and 3.0. This selection is intuitive and is based on familiarity and knowledge of maintenance teams of customer behavior as well as of gamma distribution shapes.

With the help of Equation 18.2, we can estimate the scale parameter as follows:

Corresponding to the shape factor 1.2, the scale factor is = mean/1.2 = 20/1.2 = 16.67.
Corresponding to the shape factor 2.0, the scale factor is = mean/2.0 = 20/2.0 = 10.
Corresponding to the shape factor 3.0, the scale factor is = mean/3.0 = 20/3.0 = 6.67.

Agreeing to the two customer suggestions, now the maintenance team has to predict expected variations in customer response by applying the gamma PDF.

Three sets of gamma parameters, the scale and the shape factors, set the theater for simulation. The values of μ, α, and β for the three scenarios are as follows:

Scenario I   gamma [4, 1.2, 16.67]
Scenario II   gamma [4, 2.0, 10.00]
Scenario III   gamma [4, 3.0, 6.67]

The three gamma distributions, depicting the three scenarios, are plotted in Figure 18.6.

Modes

It may be seen that each scenario has a distinctly unique mode. The modes are 7, 14, and 17 days. This means that according to Scenario I, the customer is most likely to resolve clarification queries in 7 days. According to Scenario II, the most likely clarification time is 14 days. According to Scenario III, the most likely clarification time is 17 days. These modes represent the most visible customer performance. The modes represent performance highlights.

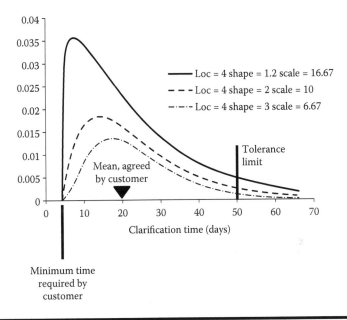

Figure 18.6 Gamma scenarios of clarification time.

Tails

In Figure 18.6, a tolerance limit is specified as 50 days. This limit marks the end of core performance and the beginning of the tail area. Area beyond the tolerance limit is the risk associated with the chosen gamma scenario.

Risks (%) in the three scenarios are 8.68, 5.40, and 3.03 from the tail areas of Figure 18.6. These risks have been computed by dividing the tail area by the total area.

BOX 18.3   GAMMA MODELING OF RAINFALL

Gamma distribution is often used in rainfall modeling.

In water resource projects, it is necessary to collect all the information related to the region and then to analyze the collected data. A frequency analysis of the rainfall data is the most commonly applied method. The hydrologist searches for a mathematical equation characterizing the available data in hand, to fill the gaps in the observations and to extrapolate it to a longer period.

Typically, two-parameter gamma distributions are fitted to rainfall data. The shape and scale parameters of the gamma distribution, α and β, are determined from the daily rainfall data of the gauging station.

Different techniques are used in estimating the parameters: the graphical method, the least squares method, the method of moments, and the maximum likelihood method.

In the analysis of 30 years of rainfall data, it is seen that α varies between 0.341 and 0.569 and β varies between 6.892 and 19.94 in a year. These gamma distribution parameters summarize the pattern of rain fall (based on Aksoy [3]).

Scenario Analysis

The gamma model enables us to evaluate three scenarios, three kinds of responses, from the customer. The first response has a shorter mode of 7 days but a higher risk of 8.68%. This could happen if the customer is requested to provide earlier response as a top priority.

The second response has a mode of 14 days but a lower risk of 5.4%. Judging by the apparent central tendency, mode, the customer seems to have slowed down, but the overall risk has reduced in a counterintuitive way.

In the third scenario, perhaps the customer is in his element, the mode is delayed further and reached a value of 17 days while the overall risk has come down further to a low value of 3.03%.

The tricky balance between demonstrated mode and real risk is a lesson we learn from this study.

Like in the case of customer clarification time in maintenance projects, gamma models can be built for internal clarification time taken by developers to respond to queries from testers in during software development. Gamma models can also be built for requirements elicitation time in software development. In all these cases, the gamma lesson can be applied:

> *Early closure is a myth; closure needs a natural time for understanding, analysis, training, and response.*

BOX 18.4   PACKING HISTORY INTO A GAMMA PDF WITH MINIMAL DATA

Rainfall data can be huge, especially when one wants to study history. Presenting descriptive statistics such as maximum, minimum, mean, median, and variance is still not adequate. Climatologists prefer to fit mathematical models such as gamma distribution to represent the overall pattern. Descriptive statistics and more are inherent in the equation. All that is required is just two parameters—the shape and scale parameters—for a season and location.

Working with a PDF such as gamma has further advantages. First, gaps in data are filled by the equation. Second, one can extrapolate mathematically and see beyond boundaries. Third, strategic planners can estimate risk.

There are several options available while choosing a PDF for rainfall data. Statistical techniques include the compound Poisson–exponential distribution; the log, square root, and cube root normal distributions; the gamma distribution; various normalizing transforms; the kappa distribution, the Weibull distribution, and the Box–Cox transformation.

However, the gamma distribution is frequently used to represent precipitation because it provides a flexible representation of a variety of distribution shapes while using only two parameters: shape and scale (based on Husak [4]).

NIST Formula for Gamma Parameter Extraction

Instead of the approach we used in plotting Figure 18.4 where we had used visual judgment of histogram of data to decide on shape and the method of moment formula given in Equation 18.2, to calculate scale, we can use the more rigorous NIST [5] formulas presented in Equations 18.5 and 18.6 for parameter extraction:

$$\text{Shape } \alpha = \left(\frac{\bar{x}}{s} \right)^2 \tag{18.5}$$

$$\text{Scale } \beta = \frac{s^2}{\bar{x}} \tag{18.6}$$

where $\bar{x}$ is the mean and s is the standard deviation.

Applying Gamma Distribution to Software Reliability Growth Modeling

Gamma distribution has been used in reliability studies. The gamma cumulative distribution function (CDF) is an S curve and can represent cumulative defects discovered in software. An example of CDF, although on a different metric, may be found in Figure 18.5. A CDF plotted with cumulative defects is known as the software reliability growth curve, a subject treated in detail in Chapter 19 where the Weibull distribution is used and in Chapter 21 where the Gompertz distribution is used.

Although the Weibull distribution is popularly used to fit failure data, the gamma distribution is more suited in certain cases. In a research study, Sonia Meskini [6] applies gamma distribution to construct a software reliability growth model for smart phones. Sonia observes,

In Skype Version 1 the Weibull distribution is the closest to the actual behavior curve of the application, followed by the gamma distribution. In Skype Version 2 it is to be noted that although the S-shaped distribution is a particular case of the gamma distribution, it fits the data slightly better. For Skype Version 3 it can be concluded that the gamma distribution is the closest to the actual behavior curve of the application failure data, followed by the Rayleigh distribution.

In her conclusion Sonia mentions,

We collected data from all over the world and divided them into different versions, and grouped them into different time periods (days, weeks, and months). Each application version failure data, when plotted in time periods, shows the same pattern: an early 'burst of failures,' likely due to the most evident defects, followed by a steep decrease in failure rate. We first tried several non-linear distributions to better fit the failure data, and after numerous experiments, we found that the observed behavior is better modeled by the Weibull or gamma distributions.

Sonia finally states that, in her observation,

It can be noted that the gamma distribution, along with its particular S-shaped case, model more frequently the failure data.

In essence, gamma distribution is attractive because of its simplicity, to perform as on Software reliability growth model. Gamma fits elegantly to processes that are not in our direct control but dependent on external factors like customer. With less data points we can do simulation using gamma distribution.

BOX 18.5 MODELING EARTHQUAKE DAMAGE WITH GAMMA DISTRIBUTION

The application of the gamma distribution to failure data has been extended to model damages produced by earthquakes. Repeated overloading of structures due to earthquake shocks cumulates damage on the hit structure. The parameter "residual ductility to collapse" is used to measure this damage. The cumulated level of deterioration can be modeled by gamma distribution. The occurrence of earthquakes is a Poisson process (see Chapter 12, Law of Rare Events), but cumulative wear is a gamma process. Structure reliability is obtained by estimating the probability of cumulated damage exceeding the threshold (based on Iervolino and Chioccarelli [7]).

Review Questions

1. What is the relationship that connects the scale and shape parameters of gamma distribution?
2. How will you estimate the scale parameter, if you can guess the shape parameter and have the value of mean of all observations?
3. How will you estimate the scale parameter directly from data?
4. Mention three applications of the gamma distribution.
5. What is the difference between a Poisson process and a gamma process?

Exercises

1. Plot a gamma curve using Excel function GAMMA.DIST for a shape of 3 and a scale of 10 units.
2. Plot a software reliability growth model using gamma distribution with a scale of 30 days and a shape of 2.2.
3. For the SGRM you have drawn in Exercise 2, calculate the fraction of defects remaining in the software on day 60. (Clue: use Excel function GAMMA.INV to calculate the result.)
4. The mean value of a certain data set is 32.2. If the shape factor is estimated as 3 by seeing the histogram of data, what would be the scale factor of the gamma distribution of the data?
5. If the mean of data = 12 and the standard deviation is 2, estimate the gamma shape and scale parameters to obtain a gamma model of data.

References

1. D. Kundu and A. Manglick, *Discriminating between the Log-normal and Gamma Distributions*, Faculty of Mathematics and Informatics, University of Passau, Germany.
2. J. F. Lawless, *Statistical Models and Methods for Lifetime Data*, Wiley, New York, 1982.
3. H. Aksoy, Use of gamma distribution in hydrological analysis, *Turkish Journal of Engineering and Environmental Sciences*, 24, 419–428, 2000.
4. G. J. Husak, J. Michaelsen and C. Funk, Use of the gamma distribution to represent monthly rainfall in Africa for drought monitoring applications, *International Journal of Climatology*, 27(7), 935–944, 2007.
5. *NIST/SEMATECH Engineering Statistics Handbook*, The National Institute of Standards and Technology (NIST) is an agency of the U.S. Department of Commerce. Available at http://www.nist.gov/itl/sed/gsg/handbook_project.cfm.
6. S. Meskini, *Reliability Models Applied to Smartphone Applications* (thesis), The School of Graduate and Postdoctoral Studies, The University of Western Ontario London, Ontario, 2013.
7. I. Iervolino and E. Chioccarelli, Gamma modeling of continuous deterioration and cumulative damage in life-cycle analysis of earthquake-resistant structures, *Proceedings of the 11th Conference on Structural Safety and Reliability*, New York, June 16–20, 2013.

It is a good idea to fit Weibull by equating the median of data to the median value of the distribution. Weibull, being a skewed curve, is best represented by median. This formula is simple.

Moments Method

The scale factor can be estimated by the method of moments. The mean value of the distribution is equated to the mean value of data. The variance of the distribution is equated to the variance of data.

Equation to the mean of the distribution is given as follows:

$$\mu = \beta\Gamma\left(1+\frac{1}{\alpha}\right) \tag{19.3}$$

Equation to the variance of the distribution is given as follows:

$$\sigma^2 = \beta^2\left[\Gamma\left(1+\frac{2}{\alpha}\right)-\Gamma^2\left(1+\frac{1}{\alpha}\right)\right] \tag{19.4}$$

MLE

A commonly accepted approach to the general problem of parameter estimation is based on the principle of maximum likelihood estimation (MLE). Moments-based estimators have been popular because of their ease of calculation, but MLEs enjoy more properties desirable for estimators. For a first-order judgment, the moments-based approach is good enough.

Parameters for Machine Availability Modeling

Nurmi and Brevik [1] studied the problem of machine availability in the enterprise area and wide area distributed computing settings using Weibull. In one of the models, they fit data to a Weibull with a shape factor of 0.49 and a scale factor of 2403. It may be noted that the shape factor is less than 1, making it sharper than the exponential function. The scale value is large on par with the median machine availability value.

BOX 19.2   WEIBULL, THE SCIENTIST

Ernst Hjalmar Waloddi Weibull (1887–1979) was a Swedish engineer, scientist, and mathematician. In 1914, while on expeditions to the Mediterranean, the Caribbean, and the Pacific Ocean on the research ship *Albatross*, Weibull wrote his first paper on the propagation of explosive waves. He developed the technique of using explosive charges to determine the type of ocean bottom sediments and their thickness. The same technique is still used today in off-shore oil exploration. In 1939, he published his paper on Weibull distribution in probability theory and statistics. In 1941, BOFORS, a Swedish arms factory, gave him a personal research professorship in Technical Physics at the Royal Institute of Technology, Stockholm.

Weibull published many papers on the strength of materials, fatigue, rupture in solids, bearings, and of course the Weibull distribution, as well as one book on fatigue analysis in 1961. In 1951, he presented his most famous paper to the American Society of Mechanical Engineers (ASME) on Weibull distribution, using seven case studies.

Weibull worked with very small samples at Pratt & Whitney Aircraft and showed early success. Dorian Shainin, a consultant for Pratt & Whitney, strongly encouraged the use of Weibull analysis. Many started believing in the Weibull distribution.

The ASME awarded Weibull their gold medal in 1972, citing him as "a pioneer in the study of fracture, fatigue, and reliability who has contributed to the literature for over thirty years. His statistical treatment of strength and life has found widespread application in engineering design."

Weibull's proudest moment came in 1978 when he received the great gold medal from the Royal Swedish Academy of Engineering Sciences, personally presented to him by King Carl XVI Gustaf of Sweden.

The Weibull distribution has proven to be invaluable for life data analysis in aerospace, automotive, electric power, nuclear power, medical, dental, electronics, and every industry.

Standard Weibull Curve

If the scale parameter is 1 and the location parameter is 0, the Weibull curve assumes a standard form shown below, which is completely controlled by the shape parameter as follows:

$$W(x) = \alpha x^{\alpha-1} e^{-x^{\alpha}} \tag{19.5}$$

We have plotted three standard Weibull curves for three different values of shape factor in Figure 19.2.

NIST defines the several useful properties for the standard Weibull distribution. The one we need to look at is the formula for median. This relationship is of

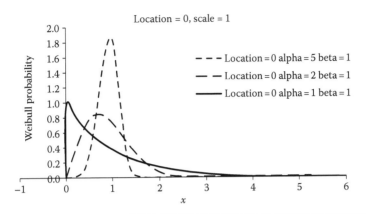

Figure 19.2 Weibull curves with different shape factors.

immense help for curve fitting. From the median value of data, we can directly calculate shape factor as follows:

$$\text{Median} = \ln(2)^{\frac{1}{\alpha}} \tag{19.6}$$

BOX 19.3 AN IMPRESSIVE RANGE

We can find several applications of the Weibull application in the literature; here are a few instances.

The strength of yarn is not a single-valued property but a statistical variable. The statistical distribution of yarn strength is usually described by the normal distribution. As an improvement, Weibull statistics was used by Shi and Hu [2] as a tool to analyze the strength of cotton yarns at different gauge lengths to find the relationship between them.

Propagation delay in CMOS circuits is characterized by Weibull distribution. The experiments of Liu et al. [3] on large industrial design demonstrate that the Weibull-based delay model is accurate, realistic, and economic.

Fire interval data are known to belong to the Weibull family. A study by Grissino-Mayer [4] shows that two- and three-parameter Weibull distributions were fit to fire interval data sets. The three-parameter models failed to provide improved fits versus the more parsimonious two-parameter models, indicating that the Weibull shift parameter may be superfluous.

Reliability can be predicted only with the help of suitable models. Sakin and Ay [5] studied reliability and plotted fatigue life distribution diagrams of glass fiber-reinforced polyester composite plates using a two-parameter

Weibull distribution function. The reliability percentage can be found easily corresponding to any stress amplitude from these diagrams.

Robert et al. [6] used the Weibull probability density function as a diameter distribution model. They stated, "Many models for diameter distributions have been proposed, but none exhibit as many desirable features as the Weibull. Simplicity of algebraic manipulations and ability to assume a variety of curve shapes should make the Weibull useful for other biological models."

Three-Parameter Weibull

We can think of a general Weibull curve by including a location parameter as follows:

$$W(x) = \frac{\alpha}{\beta}\left(\frac{x-\mu}{\beta}\right)^{\alpha-1} e^{-\left(\frac{x-\mu}{\beta}\right)^{\alpha}} \quad x > \mu,\ \alpha,\ \beta > 0 \quad (19.7)$$

where the fitted values of the parameters are as follows: μ is the location parameter, α is the shape parameter, and β is the scale parameter.

This equation has been generated by substituting x by $(x - \mu)$ in Equation 19.2.

Figure 19.3 shows three Weibull curves with three location parameters, the shape and the scale are kept constant. It is evident how changing the location parameter shifts the curve along the x-axis. When location = 0, we obtain the basic Weibull curve, which stays on the positive side of the x-axis. This by itself offers an advantage of modeling nonzero values. When location value is increased, the curve has a potential to model real life data with higher positive numbers that stay at

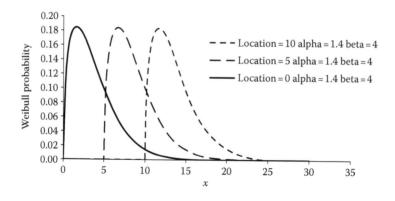

Figure 19.3 Weibull curves with different location parameters.

some distance from the origin, such as software productivity. First, we should park the curve at an optimal location and the adjusted scale and shape until we obtain minimum error in fitting.

The use of this three-parameter model is illustrated in the following paragraphs.

In his 1951 path-breaking paper "A Statistical Distribution Function of Wide Applicability," Weibull presented seven case studies of the Weibull distribution application [7]. The case for BOFORS steel strength is interesting; hence, we studied this case to illustrate the modern version of the Weibull distribution. Weibull presented steel strength data that we convert into a histogram, as shown in Figure 19.4.

We have fitted the three-parameter Equation 19.7 to Weibull's data using the following steps:

1. The location parameter was fixed at the minimum value of reported steel strength, 32 units.
2. The choice of shape parameter α is based on the shape of the histogram shown in Figure 19.4. The choice of 2 is obvious.
3. The scale parameter was adjusted to obtain the minimum least square error. We begin the iteration by keeping the initial value of the scale factor at the median departure from the minimum value. The initial value of the scale factor thus obtained is 3.5. We calculate the mean square error at this stage between the predicted and the actual probability value. Tentative perturbations of scale factor indicate that moving up reduces error. We increase the scale factor in steps of 0.1 until the minimum error is achieved. The best value of scale parameter happens to be 4.7.

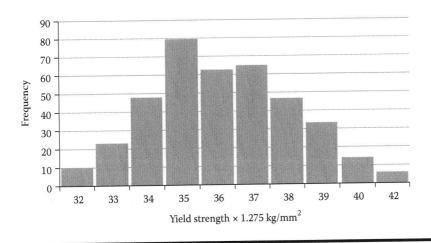

Figure 19.4 Historical steel strength data used by Weibull, the scientist.

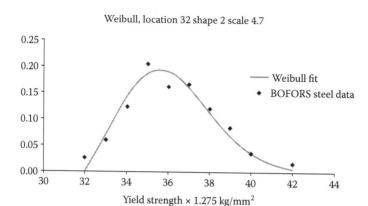

Weibull, location 32 shape 2 scale 4.7

Figure 19.5   Historical Weibull model of steel strength.

The fitted curve is shown in Figure 19.5.

The highlight of the model lies in introducing a location parameter that is necessary, in this case, where data have a large minimum value.

Software Reliability Studies

Defect discovery during the life cycle follows a Weibull curve. The curve mathematically extends to beyond the release date. The tail area depends on discovery rates before release and is governed by the Weibull equation. Figure 19.6 shows a two-parameter Weibull PDF with a shape parameter of 2 and a scale parameter of 20 days, a defect discovery curve, with release date marked; the tail beyond release denotes residual defects.

Figure 19.7 is the cumulative distribution function (CDF) of the same Weibull, but the interpretation is interestingly different.

The *y*-axis represents the percentage of defects found and is likened to product reliability at any release point. Using this model, we can predict reliability at delivery.

Both Figures 19.4 and 19.5 provide approximate but useful judgments about defect discovery.

> *Above all they provide valuable clues to answer the question, to release the product or test it further?*

An experienced manager can make objective decisions using these graphical clues. Some organizations have attempted to declare the CFD value at release date as software reliability and use this numerical value to certify the product.

Vouk [8] reported the use of Weibull models in the early testing (e.g., unit testing and integration testing phases) of a very large telecommunication system. It is

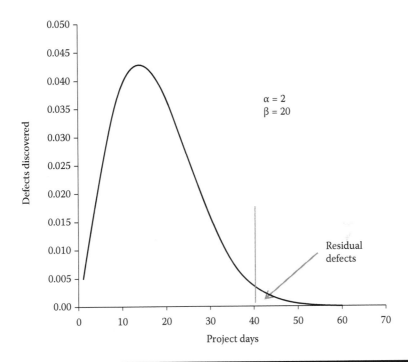

Figure 19.6   Weibull distribution—defect discovery model.

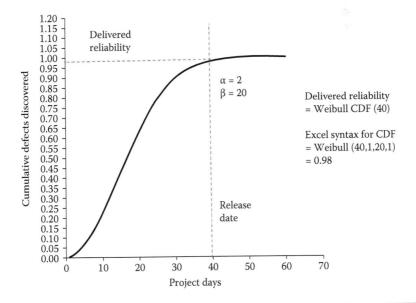

Figure 19.7   Weibull distribution CDF—delivered reliability.

shown that the dynamics of the nonoperational testing processes translates into a Weibull failure detection model. Vouk also equated the Weibull model (of the second type) into the Rayleigh model. He affirmed that the progress of the nonoperational testing process can be monitored using the cumulative failure profile. Vouk illustrated and proved that the fault removal growth offered by structured based testing can be modeled by a variant of the Rayleigh distribution, a special case of the Weibull distribution.

In a novel attempt, Joh et al. [9] used Weibull distribution to address security vulnerabilities, defined as software defects "which enables an attacker to bypass security measures." They have considered a two-parameter Weibull PDF for this purpose and built the model on the independent variable *t*, real calendar time. The Weibull model has been attempted on four operating systems. The Weibull shape parameters are not fixed at around 2, as one would expect; they have been varied from 2.3 to 8.5 in the various trials. It is interesting to see the Weibull curves generated by shapes varying from 2.3 to 8.5. In Figure 19.8, we have created Weibull curves with three shapes covering this range: 2.3, 5, and 8.5.

Tai et al. [10] presented a novel use of the three-parameter Weibull distribution for onboard preventive maintenance. Weibull is used to characterize system components' aging and age-reversal processes, separately for hardware and software. The Weibull distribution is useful "because by a proper choice of its shape parameter, an increasing, decreasing or constant failure rate distribution can be obtained." Weibull distribution not only helps to characterize the age-dependent failure rate of a system component by properly setting the shape parameter but also allows us to model the age-reversal effect from onboard preventive maintenance using "location parameter." Weibull also can handle the service age of software and the service age of host hardware. They find the flexibility of the Weibull model very valuable while making model-based decisions regarding mission reliability.

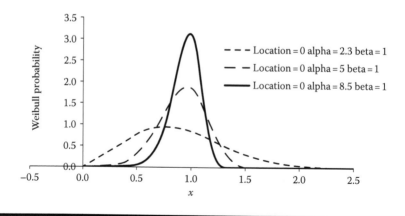

Figure 19.8 When Weibull shape factor changes from 2.3 to 8.5.

Putnam's Rayleigh Curve for Software Reliability

Remembering that the Rayleigh curve is similar to the Weibull Type II distribution, let us look at Lawrence Putnam's Rayleigh curve, as it was called, for software reliability:

$$E_m = 6 \frac{E_r}{t_d^2} te^{-3\frac{t^2}{t_d^2}}$$

(19.8)

$$\text{MTTD} = \frac{1}{E_m} \text{ after milestone } 4$$

where E_r is the total number of errors expected over the life of the project; E_m is the errors per month; t is the instantaneous elapsed time throughout the life cycle; t_d is the elapsed time at milestone 7 (the 95% reliability level), which corresponds to the development time; and MTTD is the mean time to defect.

To apply this curve software, the life cycle must be divided into nine milestones as prescribed by Putnam and define the parameters by relating them to the relevant milestones [11,12]. There are many assumptions behind this equation. This model works better with large full life cycle projects that follow the waterfall life cycle model. Putnam claimed,

> With this Rayleigh equation, a developer can project the defect rate expected over the period of a project.

Cost Model

Putnam used the same Rayleigh curve to model manpower build up and cost during the software development project. The equation was used in his estimation model. The same equation was used to define productivity. Putnam's mentor Peter Norden had originally proposed the curve in 1963. Putnam realized its power and applied it well. The equation came to be known popularly as the software equation. Technically, it is known as the Norden–Rayleigh curve.

Putnam recalls [11],

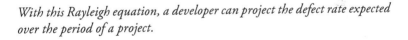

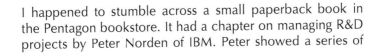

I happened to stumble across a small paperback book in the Pentagon bookstore. It had a chapter on managing R&D projects by Peter Norden of IBM. Peter showed a series of curves which I will later identify as Rayleigh curves. These curves traced out the time history of the application of people to a particular project. It showed the build up, the peaking and the tail off of the staffing levels required to get a research and development project through that process and into production. Norden pointed out that some of these projects were for software projects, some were hardware related, and some were composites of both. The thing that was striking about the function was that it had just two parameters. One parameter was the area under the curve which was proportional to the effort (and cost) applied, and the other one was the time parameter which related to the schedule.

Defect Detection by Reviews

Defect detection by reviews can be modeled by using the Weibull curve. Typically, we begin with a shape factor of 2, the equivalent of the Rayleigh curve.

Figure 19.9 shows a typical Weibull model for review defects. The cumulative curve ends at 0.86, before becoming flat. This roughly indicates that 14% of defects are yet to be uncovered, but the review process has been stopped beforehand. The PDF is clearly truncated.

At a more granular level, the review defect Weibull model can be applied to requirement review, design review, code review, and so on. It is cheaper to catch defects by review than by testing, and using such models would motivate reviewers to spend extra effort to uncover more defects.

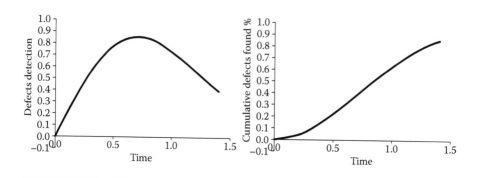

Figure 19.9 Weibull curve of family 2 showing premature closure of review.

New Trend

The defect discovery performance of contemporary projects may be seen to be dramatically different from the traditional models. In a recent data benchmark done by Bangalore SPIN (2011–2012), a consolidated view of the defect curves across eight life cycle phases is presented. A summary is available for public view in their website bspin.org, and a detailed report can be obtained from them.

BSPIN reports two types of defect curves coming from two groups of projects shown in Figure 19.10.

In one group, the unit test seems to be poorly done, and the defect curve is not a smooth Rayleigh but broken. It looks as if the first defect discovery process ends and another independent defect discovery process starts during later stages of testing. The defect curve is a mixture. As a result, the tail is fat, much fatter than the homogeneous Rayleigh (Weibull II) would be able to support. In the second group unit, the test was performed well and the defect curve had no tail. Again, the defect curve with abruptly ending slope is not the unified Putnam's Rayleigh.

We may have to look for a special three-parameter Weibull with unconventional shape factors to fit these data. Both the curves affirm that a new reality is born in defect management. There is either a low maturity performance where the testing process shows a disconnect resulting in postrelease defects or a high maturity process with effective early discovery that beats the Rayleigh tail: a mixture curve with double peak or a tailless curve. In the first case, the vision of a smooth, unified, homogeneous and disciplined defect discovery has failed. In the second case, defect discovery technology has improved by leaps and bounds, challenging the traditional Rayleigh curve.

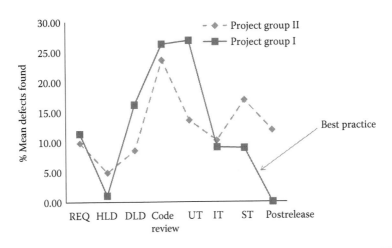

Figure 19.10 Defect distribution across project lifecycle phases.

This change reflects changes in life cycle models and practices. The modern projects follow new paradigms, self-designed and custom-tailored.

Weibull Model for Defect Prediction—Success Factors

Weibull model, be it for cost prediction or for defect prediction, is holistic in nature and can be applied early in the project. Weibull holistic models should be used by leaders to predict the future and to manage projects. With a few data and indicators, one can foresee what lies ahead.

However, there is some reluctance in applying holistic models when a lot more details have accumulated. People refuse to look at the larger picture. Weibull distribution presents a larger picture. The predictive use of Weibull cost models that resemble earned value graphs or Weibull defect curves that resemble reliability growth graphs is easily forgotten.

Alagappan [13] presented three patterns of details that overstep the holistic Rayleigh curve:

1. Fluctuating defect trend
2. Lower actual defect density
3. Effective defect management (early defect removal)

These details come after project closure. Weibull can be constructed reliably enough halfway through the project. This advantage is untapped by the industry.

Stoddard and Goldenson [14] mentioned the Rayleigh model as a process performance model in an SEI Technical Report (2010). The report presents a successful case study on a Rayleigh model from Lockheed Martin in which the authors of the case study described,

> The use of Rayleigh curve fitting to predict defect discovery (depicted as defect densities by phase) across the life cycle and to predict latent or escaping defects. Lockheed Martin's goal was to develop a model that could help predict whether a program could achieve the targets set for later phases using the results to date.
>
> Research indicates that defects across phases tend to have a Rayleigh curve pattern. Lockheed Martin verified the same phenomenon even for incremental development approaches, and therefore decided to model historical data against a Rayleigh curve.
>
> Two parameters were chosen: the life-cycle defect density (LDD) total across phases and the location of the peak of the curve (PL). Outputs included planned defect densities by phase with a performance range based on the uncertainty interval for LDD and PL and estimated latent defect density.

Inputs to the model during program execution included the actual defect densities by phase. Outputs during program execution included fitted values for phases to date, predicted values for future phases and estimated LDD, PL, and latent defect density.

The presentation suggested that those who want to be successful using defect data by phase for managing a program need to be believers in the Rayleigh curve phenomenon for defect detection.

The belief in the Rayleigh curve, mentioned as an ingredient for success, is the point we wish to highlight.

BOX 19.4 VIEWS OF NORDEN WHO FIRST APPLIED WEIBULL TO SOFTWARE ENGINEERING

Peter V. Norden has been a consultant with IBM's Management Technologies practice, specializing in the application of quantitative techniques to management problems and project management systems. He was a member of the team that developed IBM's worldwide PROFS communication system, which eventually became the Internet.

Norden [15] created history by applying the Weibull distribution to software development. He was building quantitative models when he noticed,

---

It turned out, however, that time series and other models built on these data had relatively poor predictive value. It was only when we noticed that the manpower build-up and phase-out patterns related to why the work was being done (i.e., the purpose of the effort, such as requirements planning, early design, detail design, prototyping, release to production) that useful patterns began to emerge. The shapes were related to problem-solving practices of engineering groups and explained by Weibull distributions.

Subsequent researchers (notably Colonel L. H. Putnam, originally of the U.S. Army Computer Systems Command) referred to them as Rayleigh curves but were dealing with the same phenomenon.

The life cycle equation computes the level of effort (labor-hours, labor-months, etc.; the scale is arbitrary) required in the next work period (day, week, month, etc.) as a function of the time elapsed from the start of this particular cycle, the total effort forecast for the cycle, and a scaleless "trashiness" parameter that could represent the urgency of the job.

Review Questions

1. What settings will make a Weibull curve behave like a Rayleigh distribution?
2. What is the role played by location parameter?
3. What is the formula connecting the scale shape and scale factor of Weibull curve?
4. Who invented the Weibull distribution?
5. Who applied the Weibull distribution to software projects for the first time?

Exercises

1. The median value of a certain data set is 4.5. The data are suspected to have standard Weibull distribution. Calculate the scale shape.
2. Plot a Weibull curve with a shape factor of 3 and a scale factor of 15, making use of the Excel function WEIBULL.DIST. (Clue: set the cumulative value = 0.)
3. In software review, defect discovery follows the Weibull model with a shape of 2 and a scale of 15 days. Find the remaining defects in the code if review is terminated on day 20.
4. Software productivity data (lines of code [LOC] per person day) is fitted to Weibull with the following parameters: location = 30, shape = 3, and scale = 50. Find the probability that productivity will go above 70 LOC/person-day.
5. Fit Putnam's software reliability model to BSPIN data (graphs in Figure 19.8) and predict the percentage of postrelease defect.

References

1. D. Nurmi and J. Brevik, *Modeling Machine Availability in Enterprise and Wide-Area Distributed Computing Environments*, UCSB Computer Science Technical Report Number CS2003-28.
2. F. Shi and J. Hu, The effect of test length on strength of cotton yarns, *Research Journal of Textile and Apparel*, 5(1), 18–25, 2001.
3. F. Liu, C. Kashyap and C. J. Alper, *A Delay Metric for RC Circuits Based on the Weibull Distribution*, DAC '98 Proceedings of the 35th Annual Design Automation Conference, pp. 463–468, ACM New York, 1998.
4. H. D. Grissino-Mayer, Modeling fire interval data from the American southwest with the Weibull distribution, *International Journal of Wildland Fire*, 9(1), 37–50, 1999.
5. R. Sakin and I. Ay, Statistical analysis of bending fatigue life data using Weibull distribution in glass-fiber reinforced polyester composites, *Materials and Design*, 29, 1170–1181, 2008.
6. L. Robert, T. Bailey and R. Dell, Quantifying diameter distributions with the Weibull function, *Forest Science*, 19(2), 97–104, 1973.
7. W. Weibull, A statistical distribution function of wide applicability, *Journal of Applied Mechanics*, 18, 293–297, 1951.

8. M. A. Vouk, *Using Reliability Models During Testing with Non-Operational Profiles*, North Carolina State University, 1992.
9. H. C. Joh, J. Kim and Y. K. Malaiya, Vulnerability discovery modeling using Weibull distribution, *19th International Symposium on Software Reliability Engineering*, IEEE, 2008.
10. A. T. Tai, L. Alkalai and S. N. Chau, *On-Board Preventive Maintenance: A Design-Oriented Analytic Study for Long-Life Applications*, Jet Propulsion Laboratory, California Institute of Technology.
11. L. H. Putnam and W. Myers, *Familiar Metric Management—Reliability*. Available at http://www.qsm.com/fmm_03.pdf.
12. *Larry Putnam's Interest in Software Estimating*, Version 3, Copyright Quantitative Software Management, Inc., 1996.
13. V. Alagappan, Leveraging defect prediction metrics in software program management, *International Journal of Computer Applications*, 50(20), 23–26, 2012.
14. R. W. Stoddard II and D. R. Goldenson, *Approaches to Process Performance Modeling: A Summary from the SEI Series of Workshops on CMMI High Maturity Measurement and Analysis*, SEI Technical Report, CMU/SEI-2009-TR-021 ESC-TR-2009-021, January 2010.
15. P. V. Norden, Quantitative techniques in strategic alignment, *IBM Systems Journal*, 32(1), 180–197, 1993.

Chapter 20

Gumbel Distribution for Extreme Values

A Science of Outliers

Convention has it that outliers must be marked, studied, and analyzed for root causes. In process management, outliers represent high cost, poor quality, and rework. The temptation seems to be to attach a stigma to outliers and build probability density functions (PDFs) for the remaining data. A scientific way would be to treat outliers statistically and even predict their occurrence. These outliers can be called *extreme values* and be subjected to treatment by the science of *extreme value theory*, invented by Fréchet (see Box 20.1). The behavior of extremes can be modeled by extreme value distributions.

Cláudia Neves et al. [1] summarized the characteristics of extreme value distributions as follows:

> A distribution function that belongs to the Fréchet domain of attraction is called a heavy-tailed distribution, the Weibull domain encloses light-tailed distributions with finite right endpoint and the particularly interesting case of the Gumbel domain embraces a great variety of tail distribution functions ranging from light to moderately heavy, whether detaining finite right endpoint or not.

BOX 20.1 FIVE PEOPLE AND EXTREME VALUE THEORY

Five people have contributed to extreme value theory. Fréchet proposed an extreme value distribution in 1927. Fisher and Tippet refined it in 1928 and proposed three types of extreme value distributions. In 1948, Gnedenko formulated the Fisher–Tippett–Gnedenko theory (generalized extreme value theory). Gumbel worked on type I extreme value distribution (called the Gumbel distribution after him) and provided simpler derivation and proof in 1958.

Fréchet—Maurice René Fréchet (1878–1973), a French mathematician who made several important contributions to the field of statistics and probability.

Fisher—Ronald Aylmer Fisher (1890–1962), an English statistician who created the foundations for modern statistical science.

Gumbel—Emil Julius Gumbel (1891–1966), a German mathematician and political writer who derived and analyzed the probability distribution that is now known as the Gumbel distribution in his honor.

Tippett—Leonard Henry Caleb Tippett (1902–1985), an English statistician who pioneered extreme value theory along with R. A. Fisher and Emil Gumbel.

Gnedenko—Boris Vladimirovich Gnedenko (1912–1995), a Soviet mathematician who is a leading member of the Russian school of probability theory and statistics.

Of the three types of extreme value distributions, the more popular one is the Gumbel distribution (see Box 20.4). There are different notations corresponding to the application of the Gumbel distribution. We follow the notation used in the *NIST Handbook*, where this is known as *type I extreme value distribution* [2].

The presence of extremes in process data may be seen in the box plot presentation of data (see Chapter 4). Beyond the threshold called *fences*, we can see extreme values on either end of typical box plots. On the right, we have extremes known as "maxima," and on the left, we have extremes known as "minima." Both the extremes can have a significant effect on the process. Gumbel distributions can be used to model both the maxima and the minima.

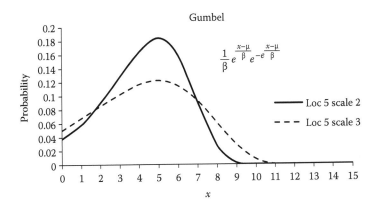

Figure 20.1 Gumbel minimum.

Gumbel Minimum PDF

Extreme minimum values follow the Gumbel distribution defined as follows:

$$F(x) = \frac{1}{\beta} e^{\frac{x-\mu}{\beta}} e^{-e^{\frac{x-\mu}{\beta}}} \qquad (20.1)$$

where μ is the location parameter and β is the scale parameter.

We have two plots of the Gumbel PDF in Figure 20.1, with a common location parameter (5) and two scale parameters (2 and 3).

It may be noted that these curves show a sharp decline in the right because they represent limits of minimal values.

BOX 20.2   WORLD RECORD: 100-METER SPRINT

In a research of world records for the 100-meter running from 1991 to 2008, extreme value theory has been applied [3] to predict the ultimate world record. Researchers predict that the best possible time that could be achieved in the near future is 9.51 seconds for men and 10.33 seconds for women. World records during the study are 9.69 and 10.49 seconds for men and women, respectively.

They used a generalized extreme value distribution, as follows:

$$G\gamma(x) = e^{-(1+\gamma x)^{\frac{1}{\gamma}}} \text{ for } 1+\gamma x \geq 0 \qquad (20.2)$$

where γ is the extreme value index.

As they studied the ultimate world records, they were interested in the right end point of the distribution. The end point is finite if $\gamma < 0$ and infinite if $\gamma > 0$. Moreover, it may be seen that in case of $\gamma < 0$, $\gamma = 0$, or $\gamma > 0$, the G_γ reduces to Weibull, Gumbel, or Fréchet distribution function, respectively. It turns out that researchers have used the reversed Weibull form of extreme value distribution.

To build the model, researchers collected the fastest personal best times. Thus, each athlete only appeared once on their list. The sample size is 762 for men and 479 for women. The estimates of γ are -0.18 for women and -0.19 for men.

The prediction is sensitive to the data window. If records up to 2005 were used, researchers find, the predictions of ultimate sprint records would be 9.29 and 10.11 seconds.

Gumbel Parameter Extraction—A Simple Approach

We can use the moments method to extract the two Gumbel parameters, location and scale. Let us consider the equations relating data mean, median and mode, and standard deviation to Gumbel parameters, shown as follows (http://en.wikipedia.org/wiki/Gumbel_distribution):

Data mode = μ
Data mean = $\mu + 0.5772\beta$
Data median = $\mu - \beta\ln(\ln(2))$
Data SD = $\beta\pi/\sqrt{6}$

Solving the previously mentioned equations will yield Gumbel parameters.

BOX 20.3   GUMBEL DISTRIBUTION, TIPPETT, AND COTTON THREAD

The evolution of the Gumbel distribution is associated with the story of cotton thread failure in the textile industry.

Leonard Henry Caleb Tippett, after graduating from Imperial College in 1923, was awarded a studentship by the British Cotton Industry Research Association (the Shirley Institute) to study statistics under Professor Karl Pearson. Later, he also worked with the great Sir Ronald Fisher.

He spent the next 40 years working at the Shirley Institute. He put statistics to work in a variety of industrial problems, such as the problem that looms in weaving sheds that were idle approximately 30% of the time, the problem of yarn breakage rates in weaving, the problem of the relationship between the length of a test specimen of a yarn and its strength, and the problem of thickness variation along the length of a yarn. He conducted factorial experiments on yarn.

The strength of the yarn is in the weakest part. This was seen by Tippett as an "extreme" situation. He studied the occurrence of extremes and identified three forms of extremes. While working with Fisher, he created the distributions, known as the Fisher–Tippet distributions. Later, Gumbel took up a special case represented by one of the three equations, simplified it, and created the Gumbel distribution.

Tippett was a role model for industrial statisticians. As a result of his work in the textile industry, he was awarded the Shewhart Medal of the American Society for Quality Control.

Gumbel Minimum: Analyzing Low CSAT Scores

Customer satisfaction (CSAT) scores were traditionally measured in a Likert scale ranging from 1 to 5. Recently, effort is being made to measure CSAT on a 0 to 10 continuous scale. The latter scale allows detailed analysis. In both scales, the problem area in CSAT lies in the minimum values, which correspond to deep dissatisfaction. The minimum values on a 0 to 10 scale follow the Gumbel distribution.

This analysis is very different from the typical control charts many plot on mean CSAT scores. The mean values are too neutral to reveal customer dissatisfaction. Preparing to plot Gumbel PDF means we collect minimum values of CSAT. This by itself is a paradigm shift in CSAT measurement.

We find the mode of the gathered minimum values and use it as the location parameter of the PDF. The scale parameter is approximately equal to 1, applying the appropriate moment equation. Thus, the model parameters are as follows:

Location = 3
Scale = 1

The Gumbel minimum PDF is constructed with these parameters and is shown in Figure 20.2.

The Gumbel PDF of CSAT is an eloquent problem statement. All the low-valued outliers in CSAT data are represented in this plot.

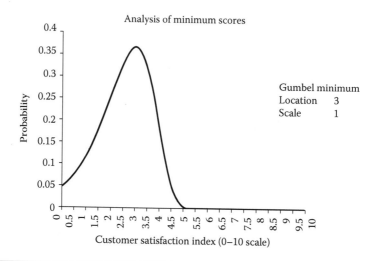

Figure 20.2 Gumbel minimum of CSAT minimum scores.

Gumbel Maximum: Complexity Analysis

Data maxima follow the Gumbel maximum PDF defined as follows:

$$F(x) = \frac{1}{\beta} e^{-\frac{x-\mu}{\beta}} e^{-e^{-\frac{x-\mu}{\beta}}}$$

(20.3)

where μ is the location parameter and β is the scale parameter.

Let us consider the case of extremely large complexity in some modules. The specification limit on cyclomatic complexity is 70. Higher values are dubbed outliers and examined one by one. We wish to use these outliers collectively as a group by constructing an exclusive PDF for these outliers. We separate these data from the database and for a special group of outliers and obtain the following statistics:

Data mean = 219.2
Data mode = 169

Thus, the model parameters, obtained by applying moments equations, are as follows:

Location = 169
Scale = 87

Using these parameters, the PDF is constructed and shown in Figure 20.3.

It shows the distribution of complexity maxima in software development. This model can be used to manage technical risks in development projects.

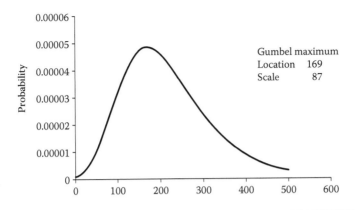

Figure 20.3 **Gumbel maximum of extreme values of cyclomatic complexity.**

The biggest problem we now have with the whole evolution of the risk is the fat-tailed problem, which is really creating very large conceptual difficulties. Because as we all know, the assumption of normality enables us to drop off the huge amount of complexity in our equation. Because once you start putting in non-normality assumptions, which is unfortunately what characterizes the real world, then these issues become extremely difficult.

Alan Greenspan (1997)

Gumbel extreme value PDF solves this problem and allows us to see risk directly and objectively instead of inadequate expressions from conventional statistical analysis.

Conventional models produce a good fit in regions where most of the data fall, potentially at the expense of the tails. In extreme value analysis, only the tail data are used.

Minima Maxima Comparisons

We proceed to compare Gumbel distributions for minima and maxima, given the same location and scale parameters. This comparison allows us to gain an insight into modeling. A comparison is illustrated in Figure 20.4.

We have kept the location parameter at 5 and scale parameter at 3 and constructed the Gumbel PDFs for minima and maxima using Equations 20.1 and 20.2.

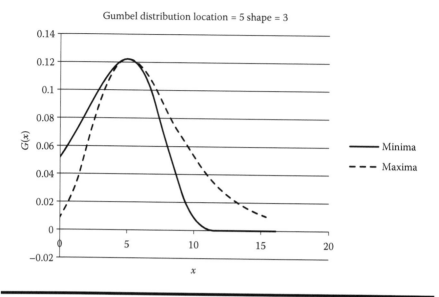

Figure 20.4 Comparison of Gumbel minimum and maximum.

BOX 20.4 HOW TO CHOOSE THE RIGHT EXTREME VALUE DISTRIBUTION

There are three types of extreme value distributions. The most common is type I, the Gumbel distribution, which is unbounded and falls off exponentially or faster. Type II, the Fréchet distribution, has a lower bound and falls off slowly according to power law and has a fat tail. This is used to model maximum values. Type III, the reversed Weibull distribution, has an upper bound and is used to model minimum values.

Type I (Gumbel)
$$G(x) = e^{-e^{-\frac{x-b}{a}}} \qquad -\infty < x < \infty \qquad (20.4)$$

Type II (Fréchet)
$$G(x) = e^{-\left(\frac{x-b}{a}\right)^{-\alpha}} \qquad x > b \qquad (20.5)$$
$$G(x) = 0 \qquad x \leq b$$

Type III (Weibull) $\quad G(x) = e^{-\left(-\left(\frac{x-b}{a}\right)\right)^{-\alpha}} \qquad x < b \qquad$ (20.6)

$\qquad\qquad\qquad\quad G(x) = 1 \qquad\qquad\qquad x \geq b$

Although the behavior of the three laws is completely different, they can be combined into a single parameterization containing one parameter ξ that controls the "heaviness" of the tail, called the *shape parameter*:

$$\text{GEV} \quad G(x) = e^{-\left[1+\xi\left(\frac{x-\mu}{\sigma}\right)\right]^{-\frac{1}{\xi}}} \qquad (20.7)$$

The location parameter μ determines where the distribution is concentrated. The scale parameter σ determines its width. The shape parameter ξ determines the rate of tail decay (the larger ξ, the heavier the tail), with the following:

$\xi > 0$ indicating the heavy-tailed (Fréchet) case.

$\xi = 0$ indicating the light-tailed (Gumbel) case.

$\xi < 0$ indicating the truncated distribution (Weibull) case.

The extreme value distributions have been differently adopted by different users. Each type of distribution offers certain advantages over the others for specific cases.

In earthquake modeling, Zimbidis et al. [4] preferred to use type III extreme value distribution (Weibull). They analyzed the annual maximum magnitude of earthquakes in Greece during the period 1966–2005. The plot of mean excess over a threshold indicates a very short tail, and researchers have chosen Weibull accordingly.

In worst-case execution time analysis of real-time embedded systems, Lu et al. [5] used the Gumbel distribution after selecting the data very carefully using special sampling techniques. Their predictions agree closely with observed data.

However, in the probabilistic minimum interarrival time analysis of embedded systems, Maxim et al. [6] found that the Weibull extreme value distribution fits better.

During the analysis of wave data, Caires [7] found the general extreme value model more suitable.

The choice depends on data.

Analyzing Extreme Problems

Instead of seeing problems as tails of some parent distribution, extreme value analysis using Gumbel distributions allows us to look at the problem squarely in the eye.

The Gumbel distribution can be used to analyze several extreme problems in addition to the two we have discussed so far. For example, we can do a simple schedule variance analysis by collecting data, as shown in Figure 20.5.

This shows the distribution of maximum values of schedule variances in a development project. The PDF is built with a location parameter of 20 and a scale parameter of 12.

Likewise, we can easily analyze extremely error prone modules, extremely costly effort escalations, extreme volatility of requirements, and so on. There is a great opportunity for such modern and innovative analysis.

There are a few cautions to be taken before we do extreme value analysis.

First, data collection needs care. Data must be drawn from samples that are independent and identical (the iid criterion). Extreme values in a single organization approximately meet this requirement of identicality, assuming similar process run in all projects. Data samples also can be easily made independent (one sample does not influence another). Doing extreme value analysis across distinctly different processes is not suggested.

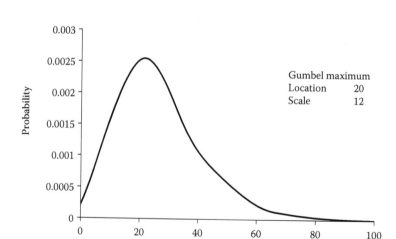

Figure 20.5   Gumbel maximum of extreme values of schedule variance.

BOX 20.5 LIVING ON THE EDGE

Combine the extremes, and you will have the true center.

Karl Wilhelm Friedrich Schlegel

Which one captures the essential feature of life: central tendency, dispersion, or extreme values? Extreme values have crossed the boundary; they contain novelty and throb with energy. They have been pushed by extreme circumstances and cruel experiments by nature. We can learn infinitely more and gain infinitely more from extreme data than from regular data.

Life is in the boundary.

We have lived out of central limit theorem, and now we should reach out and consider extreme value theorem. Both theorems have been designed to tell us about limiting performances. Let us look at some of the problems modeled by extreme value theory since Tippett and Gumbel: extreme forest fire, extremely high flood levels, extreme heights of waves, earthquake, extreme heat, extreme cold, extreme loads on aircraft structure, excessive stock movements, extreme load on wind turbines, and the list is growing. All these extreme value studies aim to save lives or property.

To solve a problem, look at the problem.

However, the optimism of Extreme Value Theory (EVT) is appropriate but also exaggerated sometimes. Yet it holds great promise. The potential of EVT remains latent, much so in software engineering practices.

Review Questions

1. When should we use the Gumbel minimum distribution?
2. When should we use the Gumbel maximum distribution?
3. What is the primary use of extreme value theory?
4. Relate outliers in a box plot to Gumbel distributions.
5. Compare the bell curve with the Gumbel distribution.

Exercises

1. Program Gumbel minimum distribution in Excel. (There is no readily available function in Excel.)
2. Program Gumbel maximum distribution in Excel. (There is no readily available function in Excel.)
3. In the CSAT Gumbel model with location = 3 and scale = 1, find the risk of CSAT score falling below 2. Make use the program you have developed for exercise 1.
4. In the complexity Gumbel maximum model with location = 169 and scale = 87, find the risk of complexity exceeding 300. Make use of the program you have developed for exercise 1.
5. For schedule variance Gumbel maximum model with location parameter = 20 and scale parameter = 12, find the risk of schedule variance exceeding 40.

 ## References

1. C. Neves and M. I. Fraga Alves, Testing extreme value conditions—An overview and recent approaches, *REVSTAT: Statistical Journal*, 6(1), 83–100, 2008.
2. *Engineering Statistics Handbook*, NIST. http://www.itl.nist.gov/div898/handbook/.
3. J. H. J. Einmahl and S. G. W. R. Smeets, *Ultimate 100m World Records Through Extreme-Value Theory*, Tilburg University, AZL, Heerlen, July 10, 2009, ISSN 0924-7815.
4. A. A. Zimbidis et al., Modeling earthquake risk via extreme value theory and pricing the respective catastrophe bonds, *Astin Bulletin*, 37(1), 163–183. doi: 10.2143/AST.37.1.2020804© 2007.
5. Y. Lu, T. Nolt, I. Bate and L. Cucu-Grosjean, *A Trace-Based Statistical Worst-Case Execution Time Analysis of Component-Based Real-Time Embedded Systems*. IEEE, Emerging Technologies & Factory Automation (ETFA), 2011 IEEE 16th Conference, pp. 1–4, Toulouse, 2011.
6. C. Maxim, A. Gogonel, D. Maxim and L. Cucu-Grosjean, *Estimation of Probabilistic Minimum Inter-Arrival Times Using Extreme Value Theory*, INRIA Nancy-Grand Est, France. https://hal.inria.fr/hal-00766063/PDF/JW_2012_submitted_26september_v2.pdf.
7. S. Caires, *Extreme Value Analysis: Wave Data*, 2011. http://www.jodc.go.jp/info/ioc_doc/JCOMM_Tech/JCOMM-TR-057.pdf.

Chapter 21

Gompertz Software Reliability Growth Model

S Curves

The S curves in software project management represent cost evolution and reliability growth within projects. The strength of the S curve is that it is based on nature. It represents a natural law of growth. The forecasting power of the S curve is due to the basic concept of limiting resources that lies at the basis of any growth process.

The earliest known S curve is the logistic function, introduced to describe the self-limiting growth of a population by Verhulst in 1838, based on the *Principle of Population* published by Thomas Malthus. Benjamin Gompertz developed the Malthusian growth model further and invented the law of mortality, a Gompertz curve, in 1825 (see Box 21.1 for a brief biography).

Since then, S-shaped curves have been applied for studies in various fields. The more precise the data and the bigger the section of the S curve they cover, the more accurately the parameters can be recovered. When a system is closer to the end of its evolution, it increases the accuracy of the forecast with the S curve [1].

Modeling Reliability Growth with Gompertzian S Curves

The application of the Gompertz curve as a software reliability model (more than a century later after Gompertz introduced his curve) is of interest to us.

The Gompertz growth curve is shown in Figure 21.1. The y axis represents reliability growth. In practice, the y axis is calibrated in terms of cumulative defects

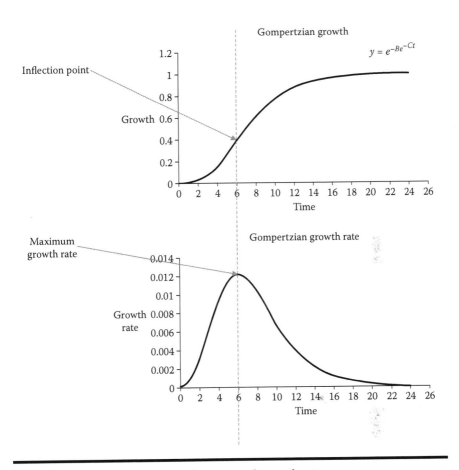

Figure 21.1   **Gompertzian growth curve and growth rate curve.**

discovered until now. We can also calibrate the y axis in terms of the percentage of defects found, if we have an estimate of the total defects in the application. The expression "percentage of defects found" is believed to be a direct measure of the percentage of reliability.

The growth curve starts at some fixed point, and the growth rate increases monotonically to reach an inflection point. After this point, the growth rate approaches a final value asymptotically. According to this model assumption, failure detection rate increases until the inflection point and then decreases.

In Figure 21.1, the inflection point is marked on the growth curve. Also shown is the growth rate curve, which is obtained by differentiating the growth curve. The inflection point in the growth curve corresponds with the peak of the growth rate curve. A vertical dashed line is drawn to show the correspondence.

Building a Reliability Growth Curve

Building a reliability growth curve in real life is fraught with basic problems. Software reliability models have not delivered the desirable deliverables that they are intended to realize, as commented by Faqih [2], who listed eight major causes for this problem,

> Unfounded types of assumption, complexity of the software, complexity of the reliability models, weakness of the reliability models, the misconception of fault and failure phenomena, inaccurate modeling parameters, difficulty in selecting the reliability model, difficulty in building software operational profile.

There are many who will adopt a cautious approach to reliability growth models. Stringfellow and Andrews [3] saw many challenges,

> It is difficult to define operational profiles and perform operational tests. Defects may not be repaired immediately. Defect repair may introduce new defects. New code is frequently introduced during the test period. Failures are often reported by many groups with different failure-finding efficiency. Some tests are more or less likely to cause failures than others. Test effort varies due to vacations and holidays.

Whether growth is Gompertzian is the next question.

Gompertz Software Reliability Growth Model Curves

The Gompertz software reliability growth model (SRGM) is defined as follows:

$$G(t) = A\left(B^{(C^t)}\right) \tag{21.1}$$

where $G(t)$ is the defects found until time (t), A is the total defects in the application, B and C are the constants with fractions between 0 and 1, and t is the time elapsed from the start of testing.

Constraining C to values less than 1 is of paramount importance; that is what makes the expression within the inner brackets decrease with time. Likewise, constraining B

to values less than 1 is equally important to make the overall function grow with time. The constant *A* is equal to the number of total defects in the application.

Figure 21.2 contains two sets of Gompertz curves. The first set of three curves are constructed with *C* = 0.3. The second set of three curves are plotted with *C* = 0.6. The value of *A* is fixed at 100.

The effect of *C* on the curves can be easily seen on the patterns that discriminate the two sets. Increasing *C* delays defect discovery.

Within each group, three values of *B* are shown: 0.1, 0.01, and 0.001. Decreasing *B* delays defect discovery.

The *x* axis of the graph is time varying from 0 to 10 units. It could be days, weeks, or months. It could also be proxies for time. However, we had weeks in mind while plotting the graph; this roughly coincides with the time required to find 100 defects in a small project.

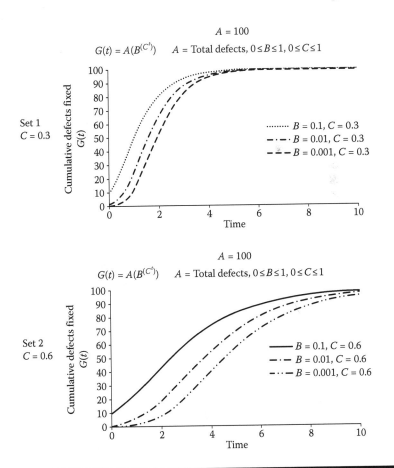

Figure 21.2 Gompertz reliability growth curves for *C* = 0.3 and *C* = 0.6.

Both B and C control the inflection point. At the inflection point, growth attains 0.36788 (this number $= \dfrac{1}{e}$, where e is the Euler constant) of the plateau value A. This means 36.788% of the total defects at the inflection point. This property characterizes the Gompertzian way of testing and finding defects. The inflection point, as we have already seen, also represents peak value in growth rate (here defect discovery rate).

The time when the inflection happens has been defined by Kececioglu et al. [4]. For the different values of B and C used in Figure 21.1, the following inflection times have been computed:

B	0.1	0.01	0.001	0.1	0.01	0.001
C	0.3	0.3	0.3	0.6	0.6	0.6
Inflection time	0.693	1.268	1.605	1.633	2.990	3.783

From these results, we may see the influence of B and C on inflection time. Large inflection time indicates delayed defect discovery.

Dimitri Shift

Kececioglu et al. [4] worked out a modified Gompertz curve by adding a constant D to displace the curve vertically by a distance D. This shift results in a four-parameter Gompertz model defined in the following equation:

$$G(t) = D + A\left(B^{(C^t)}\right) \tag{21.2}$$

This shift of the Gompertz curve in the y axis is analogous to the familiar location shift in the x axis we have seen in Chapter 20. The shift factor D is some kind of a location parameter. A plot of the shifted Gompertz curve is shown in Figure 21.3.

The shifted curve suggests a significant amount of defects discovery immediately after the start. This is not agreeable to intuitive reasoning. However, Dimitri et al. [4] claimed that the modified model fits better with data. They observe from several data sets that

> Reliability growth data could not be adequately portrayed by the conventional Gompertz model. They point-out that the reason is due to the model's fixed value of reliability at its inflection point. As a result, only a small fraction of reliability growth datasets following an S-shaped pattern could be fitted.

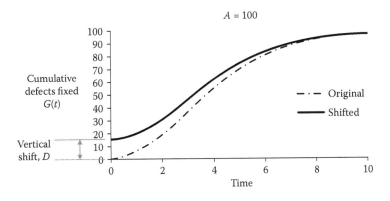

Figure 21.3   Modified Gompertz reliability growth curve with shift.

In any case, the four-parameter model offers more options during curve fitting, and this could be an advantage.

Predicting with the Gompertz Model

Once test data arrives, we may wish to begin asking the following questions:

> *How many more defects remain in the application?*
> *How long would it take to detect those defects?*

These are the prediction questions the Gompertz model strives to answer. There are two prediction scenarios. The first is when we use an auxiliary model to estimate defects in the application. For example, we might predict defects based on the size and complexity of the application using regression equations based on historical data. In this case, we know the constant A. All we have to do is to derive the remaining parameters B, C, and D from data and predict time.

In the second scenario, all four parameters are unknown. We do not have any estimate of the defects in the application. In this case, one short cut is to wait for the defect discovery rate to go through a peak and start symptoms of steady decline. The peak is the Gompertzian inflection point. If the defects found until the inflection point is known, then the overall defects A can be obtained by using the following relationship:

Defects found until inflection = total defects in the application × 0.36788

This calculation completes the prediction of total defects in the application.

We can now estimate constants B, C, and D by iterative analysis to arrive at the least square error or any other curve-fitting technique. Once these three constants are known, we can predict the time taken to achieve a given percentage of reliability. Curve fitting techniques do not assume Gompertzian behavior but force fit the Gompertz curve to data. The coefficient of determination R^2 or any other assessment of error in the fitted curve should be used to determine the quality of predictions.

Dimitri has proposed a way to extract parameters. Data are divided into three groups with an equal number of values. He proposed formulas to determine constants [4].

BOX 21.2   GOMPERTZ CURVE FOR GROWTH OF SHARKS

Sharks are the top predators and play important roles in marine ecosystems. Annual yields of small sharks in Taiwan declined dramatically from 5699 tons in 1993 to 510 tons in 2008, which implies that these stocks, mainly caught by trawlers and long-liners in coastal waters off Taiwan, have experienced heavy exploitation in recent years.

The blacktip sawtail catshark is a small species that inhabits tropical and subtropical coastal waters of the western Pacific region. In Taiwan, this species is found in coastal waters of western and northern Taiwan and is one of the most important small shark species. The growth pattern of this species has been studied by Liu et al. [5].

The growth data of 275 female sharks have been fitted to growth equations such as Gompertz, as shown in Figure 21.4.

$$L_t = A\,e^{Be^{Ct}}$$

Where the parameters have been estimated as
L_t = Length in cm at time t years
A = Asymptotic length 52.8 cms
$B = -2.28$
$C = -0.232$

The above Gompertz equation is an adaptation by Kwang-Ming Liu et al. The constants are differently limited.

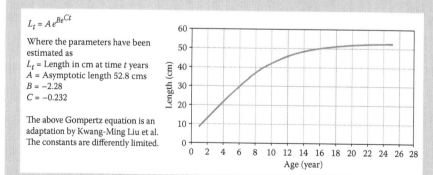

Figure 21.4   Gompertz curve for shark growth.

More Attempts on Gompertzian Software Reliability Growth Model (SRGM)

The Gompertz curve is easily one of the widely used models. There are several examples of application of Gompertz to predict software reliability. There have been different adaptations, and different interpretations, each adding to the insight into the Gompertz curve.

The following interpretations of the Gompertz equation by its different users and the values of model constants are noteworthy.

Stringfellow and Andrews

Stringfellow and Andrews [3] built a Gompertzian SRGM shown in Figure 21.5. They fitted a model with $B = 0.001$ and $C = 0.74$. The model was built from failure data collected from a large medical record system, consisting of 188 software components. The low value of B suggests delayed discovery and initially slow progress in testing.

The criteria used by Stringfellow and Andrews in model evaluation are simple and effective:

Curve fit: How well a curve fits is given by a Goel–Okumoto (GO) F test: the R^2 value.

Prediction stability: A prediction is stable if the prediction in a given week is within 10% of the prediction in the previous week.

Predictive ability: Error of predictive ability is measured in terms of error (estimate–actual) and relative error (error/actual).

Stringfellow and Andrews noted that "Gompertz performed better for Release 1 but not for Release 2 and Release 3."

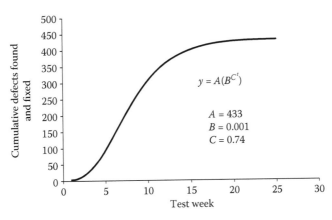

Figure 21.5   Stringfellow's version of Gompertz reliability growth model.

Zeide

Bores Zeide [6] observed,

> Another characteristic feature of the Gompertz equation is that the position of the inflection point is controlled by only one parameter, asymptotic size, A.

Swamydoss and Kadhar Nawaz

Swamydoss and Kadhar Nawaz [7] provided a physical interpretation to the parameters, as follows:

A: initially, A is taken as total defect detected until date
B: the rate at which defect rate decreases, or test case efficiency (0.27, in his study)
C: a constant, a shape parameter

Swamydoss and Kadhar Nawaz report that they have collected cumulative failures found every week and constructed the Gompertz model by curve fitting and used the model to predict reliability. If the predicted reliability values were less than the threshold, they would continue system testing.

Arif et al.

In the adaptation of Arif et al. [8], the coefficients B and C are negative numbers, not fractions.

Anjum et al.

Anjum et al. [9] published a Gompertz model with A = 191.787, B = 0.242, and C = 0.05972.

Bäumer and Seidler

Bäumer and Seidler [10] reported poor performance of Gompertz.

Ohishi et al.

Ohishi et al. [11] used the Gompertz distribution with a different mathematical structure. The equation, fitted to data, and a plot of the equation are shown in Figure 21.6.

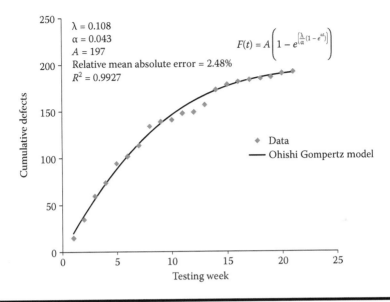

Figure 21.6 Ohishi's version of Gompertz reliability growth model.

The model predicts current reliability as 0.975 and suggests the presence of five residual defects. The model indicates that another 4 weeks of testing is required to reach a reliability level of 99%.

How to Implement Gompertz Model in Software Testing

To begin with, one must acknowledge cautions suggested by Faqih in implementing any SRGM. The concerns of Stringfellow and Andrews are also valid. These ideas have been already cited in this chapter in the Building a Reliability Growth Curve section.

Further care and preparations are required for implementing Gompertz SRGM in real life.

Gompertz is an organic model, a characteristic well seen in the way it fits the growth of sharks. Gompertz, the discoverer of the model, uses the organic force of life to drive the equation to predict mortality. Testing must be performed in an organic manner. That means that testing must be a well-coordinated and homogeneous process. There must be a visibly great understanding between testers and developers. The arrival of components for tests must follow a systematic pattern without ad hoc breaks. Testing must be performed with great sincerity. Defects must be logged promptly without delay.

Often there is an uncertainty in the exact time of defect discovery. The logged time could differ from the discovered time. The authenticity of time stamp is questionable. This uncertainty will distort data and introduce unnatural ripples into data.

To make the model more effective, test cases must be developed early, as soon as requirements are clear. All possible usage scenarios should be identified and considered during requirements collection. Requirement coverage and code coverage must be high enough.

To get better Gompertz fit, the test processes must be streamlined. Better Gompertz fit is not merely the result of statistical manipulation but presupposes testing process improvement.

Swaminathan [12] observed that proper test management and cooperation among teams are critical to the success of the Gompertz model. Some customers insist on Gompertzian testing; they expect establishment of the Gompertz model as part of the testing process and use it for prediction in an appropriate manner. Swaminathan noted,

> Customers are expecting software vendors to give an assurance on the quality of the delivered product. They insist that this assurance be backed up by a valid statistical model and mere verbal assurances would not do.
>
> How do we give this assurance to the customer?
>
> It is not just for the customers. Even for the software vendors, who have to make decisions on the size of the team for the support phase, they need to know by when the software would reach a certain maturity and thereby when they could reduce/stop testing.
>
> How can software vendors know when their product will reach a 90% maturity or 99% maturity?
>
> This has become an important aspect to address for both the customers and the vendors.

On his experience with building a live Gompertz model and running it, Swaminathan indicated,

> We started building a tool based on excel.4. The input to the tool was the number of defects found every week and the output was the predicted defects for the future weeks and the maximum number of defects that were likely to be found.
>
> Making the excel tool as per the details given was relatively simple. But the bigger challenge was to make it work for a particular team context.
>
> As the test data started flowing from the team, we started using the data to predict the maximum number of defects for the product under test.

The tools predictions turned out to be different from the intuitive predictions of the software team. When we tried to understand the reason, we found that the testing was done quite randomly.

Table 21.1 gives an example. Defects found during six weeks of testing are tabulated. Test cases executed have been included in the table to set the context and environment. Figure 21.7 contains the cumulative plot of defects. Apparently, the curve is not Gompertzian. There is a hike in the number of test cases executed that corresponds to the hike in the defects discovered.

In the tool, we were looking at only the week number and number of defects. We had ignored very crucial data in number of test cases executed. So, instead of looking at the absolute number of defects, we started looking at the defects in the context of the test cases executed every week. Even after this change, the predictions did not match with

Table 21.1　Test Results

Week No.	No. of Test Cases Executed	No. of Defects Found	Cumulative Defects Found
1	100	10	10
2	10	0	10
3	50	3	13
4	75	0	13
5	30	2	15
6	100	5	20

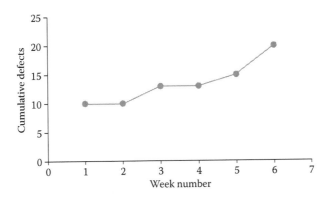

Figure 21.7　Cumulative defects found in the course of testing.

what the Project Manager expected. When we investigated the reason, it turned out that the software was not completely developed.

We learned that the Gompertz model can only be used in the test–analyze–fix cycle and not when full-fledged software development is in progress. We also learned that the model should not be used for prediction at the early stages of the test–analyze–fix cycle. The product under test has to reach a certain maturity before we start using the data for predicting.

As a Gompertzian rule of thumb, until inflection is visible, the model should not be used for prediction. It takes time and sufficient data to reach the inflection point. The model is to be kept dormant until such a time.

Gompertz Curve versus GO NHPP Model

The two-parameter GO model is discussed in Chapter 12. This original model has been criticized because it is concave and lacks the flexibility of the S curve. A third parameter c was added by Goel to this model to make it reflect real-life testing. The model is known as the *generalized GO model*, (NHPP: non homogeneous poisson process) as shown in Figure 21.8.

This model can be compared with the Gompertz curve for ready reference, also shown in Figure 21.8. The two curves shown are representative samples from the generalized GO family and the Gompertz family.

Both are S curves; the similarity is interesting coming from two different domains, one from computer science and the other from actuarial studies. Both the models, generalized GO and Gompertz curve, are capable controlled inflection. However, the Gompertz curve is a wee bit more practical and flexible.

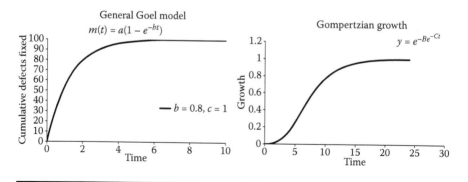

Figure 21.8   Gompertz curve versus GO NHPP model.

Review Questions

1. On what principle is the Gompertz curve grounded?
2. How many parameters control the original Gompertz curve?
3. How many parameters control the modified Gompertz curve?
4. Compare the Gompertz curve with the generalized GO SRGM.
5. What precautions must be taken before implementing the Gompertz SRGM?

Exercises

1. Calculate the inflection point reliability for a Gompertz curve with $A = 123$.
2. Download Dimitry Kucharavy and De Guio's [1] paper and find the formula for inflection time in Equation 20. Using this formula, calculate inflection time if $A = 123$, $B = 0.01$, and $C = 0.5$.
3. Predict the remaining defects in an application if the testing process has just crossed the peak discovery rate and at the inflection point 25 defects have been discovered.

References

1. D. Kucharavy and R. De Guio, Application of S shaped curves, *TRIZ—Future Conference 2007: Current Scientific and Industrial Reality*, Frankfurt, Germany, 2007.
2. K. M. S. Faqih, What is hampering the performance of software reliability models? A literature review, *Proceedings of the International MultiConference of Engineers and Computer Scientists*, Vol. I, IMECS 2009, Hong Kong, March 18–20, 2009.
3. C. Stringfellow and A. Amschler Andrews, An Empirical Method for Selecting Software Reliability Growth Models, *Empirical Software Engineering*, 7(4), 319–343, 2002.
4. D. Kececioglu, S. Jiang and P. Vassiliou, The modified Gompertz reliability growth model, *Proceedings Annual Reliability and Maintainability Symposium*, 1994.
5. K.-M. Liu, C.-P. Lin, S.-J. Joung and S.-B. Wang, Age and growth estimates of the blacktip sawtail catshark *Galeus sauteri* in northeastern waters of Taiwan, *Zoological Studies*, 50(3), 284–295, 2011.
6. B. Zeide, Analysis of growth equations, *Forest Science*, 39(3), 594–616, 1993.
7. D. Swamydoss and G. M. Kadhar Nawaz, An enhanced method of LMS parameter estimation for software reliability model, *International Journal of Innovative Research in Science, Engineering and Technology*, 2(7), 2667–2675, 2013.
8. S. Arif, I. Khalil and S. Olariu, On a versatile stochastic growth model, *Mathematical Biosciences and Engineering*. Available at http://www.cs.odu.edu/~sarif/Growth-model-MBE-rev-10-12-11.pdf, 2009.
9. M. Anjum, Md. A. Haque and N. Ahmad, Analysis and ranking of software reliability models based on weighted criteria value, *International Journal of Information Technology and Computer Science*, 5(2), 1–14, 2013.

10. M. Bäumer and P. Seidler, *Predicting Fault Inflow in Highly Iterative Software Development Processes* (Master thesis on Software Engineering), Blekinge Institute of Technology, Sweden, February 2008.
11. K. Ohishi, H. Okamura and T. Dohi, Gompertz software reliability model: Estimation algorithm and empirical validation, *Journal of Systems and Software*, 82(3), 535–543, 2008.
12. Swaminathan, *Practical Considerations While Implementing Gompertz Model* (private communications), 2013.

Suggested Readings

Davis, J., *An Applied Spreadsheet Approach to Estimating Software Reliability Growth Using the Gompertz Growth Model*. Available at http://smaplab.ri.uah/ice/davis.pdf (accessed September 25, 2002).

Dubeau, F. and Y. Mir, *Growth Models with Oblique Asymptote*, Université de Sherbrooke 2500 Boulevard de l'Université Sherbrooke (Qc), Canada, J1K 2R, Paper published in *Mathematical Modelling and Analysis*, 18, 204–218, 2013.

Index

Page numbers followed by f, t, b, and d indicate figures, tables, boxes, and data, respectively.